SpringerBriefs in Anthropology

Anthropology and Ethics

Series Editor

Richard J. Chacon, Department of Sociology and Anthropology, Winthrop University, Rock Hill, SC, USA

SpringerBriefs in Anthropology and Ethics explores the many ethical ramifications of conducting various types of anthropological research. This SpringerBrief series provides a forum where scholars (by way of case studies) can briefly address an array of ethical issues involving anthropological investigations and/or activities. The goal of the SB series is to promote comportment and investigative protocols which honor the ethical obligations that anthropologists have towards their colleagues, the discipline, and to their study populations. The series seeks to foster honest, respectful, and scholarly dialogue on a topic that has proven to be very contentious.

Maria Sapignoli • Robert K. Hitchcock

People, Parks, and Power

The Ethics of Conservation-Related Resettlement

 Springer

Maria Sapignoli
Department of Philosophy
University of Milan
Milan, Italy

Robert K. Hitchcock
Department of Anthropology
University of New Mexico
Albuquerque, NM, USA

ISSN 2195-0806 ISSN 2195-0814 (electronic)
SpringerBriefs in Anthropology
ISSN 2195-0822 ISSN 2195-0830 (electronic)
Anthropology and Ethics
ISBN 978-3-031-39266-5 ISBN 978-3-031-39268-9 (eBook)
https://doi.org/10.1007/978-3-031-39268-9

This Springer imprint is published by the registered company Springer Nature Switzerland AG
The registered company address is: Gewerbestrasse 11, 6330 Cham, Switzerland

Paper in this product is recyclable.

Foreword

People, Parks, and Power: The Ethics of Conservation-related Resettlement documents the ongoing expulsions of African, South American, and Asian indigenous populations from their ancestral homelands ostensibly to "protect the environment." Maria Sapignoli and Robert Hitchcock analyze, by way of several case studies, how various conservation-inspired measures impact indigenous and other populations in selected regions. Specifically, the authors examine the effects of relocation and resettlement events occurring in the Central Kalahari Game Reserve (Botswana), Etosha National Park (Namibia), Hwange National Park (Zimbabwe), and the Kgalagadi Transfrontier Park (South Africa and Botswana). They assess comparative material drawn from Central Africa (Batwa), Latin America (Amazonia and Gran Chaco), East and South Asia, and North America.

This work comes to us at a most auspicious moment since there are at least 260,000 officially protected national parks and other protected areas in the world with more being added every day (Deguignet et al., 2014; Dowie, 2009; International Union for the Conservation of Nature and Natural Resources, 2018; Juffie-Bignoli et al., 2014). Tragically, many of these reserves are based on the Yellowstone Conservation Model which calls for the removal of local populations so as to "protect nature." In Africa alone, Yellowstone-inspired conservation protocols have resulted in the displacement of an estimated 39 million people in order to "protect nature" (DeGeorges & Reilly, 2008; Dowie, 2009; Sapignoli et al., 2017). Likewise, vast numbers of conservation refugees have appeared throughout Asia as the result of this predatory form of conservation (Dowie, 2009).[1] In response to such gross violations of human rights, the authors argue for a community-based conservation alternative over exclusionary Yellowstone Conservation Model ("Fortress Conservation") schemes. In order to understand how and why this coercive form of conservation gained such widespread acceptance, a brief historical overview is in order.

Origins of the Yellowstone Conservation Model (Fortress Conservation)

Unprecedented population growth along with unbridled natural resource utilization in the Western world resulted in high levels of environmental degradation (Borgerhoff & Coppolillo, 2005). Troubled by this trend, in 1864, George Perkins Marsh published his *Man and Nature*, which claimed that an ecological balance existed in the environment until the arrival of human beings whose presence invariably upset this equilibrium. According to Marsh (2012: 19 [1864]), "...man is everywhere a disturbing agent. Wherever he plants his foot, the harmonies of nature are turned to discords. The proportions and accommodations of which insured the stability of existing arrangements are overthrown." For Marsh, the way to re-establish environmental equilibrium was simple: "Natural arrangements, once disturbed by man, are not restored until he retires from the field, and leaves free scope to spontaneous regenerative energies; the wounds he inflicts are not healed until he withdraws the arm that gave the blow" (Marsh, 2012: 21 [1864]). Marsh's *Man and Nature* inspired nineteenth-century conservationists to call for the establishment of "wilderness areas."[2] Thus, in these uninhabited localities, nature would be protected from the harmful effects of human settlement (except for a small number of park rangers and other personnel needed to maintain reserves) (Spence, 1999). On March 1, 1872, these efforts paid off when President Ulysses S. Grant signed the *Yellowstone National Park Protection Act* which removed more than two million acres from the public domain in order to preserve the region's "natural condition" (Spence, 1999: 39).

This governmental action conveniently ignored the fact that at the time of the park's creation, the Yellowstone area was frequented by several Native American groups (Merchant, 2007; Spence, 1999).[3] In actuality, "[t]he economies and cultures of Crow, Bannock, Shoshone, Salish, Nez Perce, and Northern Paiute *depended* on seasonal visitation to Yellowstone high country for hunting, harvesting the dietary staple camas, lodgepole cutting, and ceremonial or interindigenous gatherings" (Merchant, 2007: 165, emphasis added).[4] In effect, the 1872 *Yellowstone National Park Protection Act* converted tribal lands into a "pleasuring ground for the people, a preserved landscape which had not known the hand of humanity" (Kantor, 2007: 46).

Moreover, in the newly established park, officials believed that the presence of Native Americans in Yellowstone not only threatened the local environment but that indigenous peoples could never become "civilized" as long as they continued inhabiting such a remote location. Following a series of "Indian troubles," in 1877, Yellowstone Park officials coordinated with the War Department to exclude Native Americans from the preserve.[5] By 1886, the US Army had taken over the complete management of Yellowstone and would go on to use its military power to prevent indigenous peoples from re-entering the national park (Spence, 1999).[6] In short, President Grant's 1872 declaration transformed the Bannock, Shoshone, and other Native American groups who had foraged in the Yellowstone region for centuries into trespassers (Treuer, 2021).

Many conservationists invoked racist arguments to justify the dispossession of Native American ancestral homelands. For example, the avid environmentalist John Muir felt contempt toward the Native Americans he encountered while traveling in the mountains of California and claimed they had no place in the Sierra landscape (Sahagun, 2014). Moreover, he considered California Indians to be "dirty" and "lazy" (Muir, 1916: 226). For Muir, "[t]he true ownership of the wilderness belongs in the highest to those who love it most" (Muir cited in Spence, 1996: 27).[7]

Informed by such notions, the exclusionary Yellowstone Conservation Model ("Fortress Conservation") served as the template for the subsequent creation and management of national parks throughout the USA. The result was that Native Americans around the country were systematically "expelled from lands they had long inhabited and ranged to create recreational resources for whites...Wilderness was redefined as untainted by human presence, and parks were conceptualized as places where white tourists could be inspired by the sublimity of depopulated natural beauty" (Merchant, 2007: 163). The concept of unpopulated wilderness areas provided cover for usurping of indigenous homelands. As Kay points out, "[t]here was no [unpopulated] wilderness. In fact, the idea that North America was a 'wilderness' untouched by the hand of man before 1492 is a myth, a myth that may have been created, in part, to justify appropriation of aboriginal lands and the genocide that befell native peoples" (1998: 490).

Along these lines, Kantor (2007: 42) argues that uninhabited parks "are built upon an illusion. They seem to offer us a rare chance to experience the continent as it was, to set eyes on a vista unspoiled by human activity. This uninhabited nature is a recent construction. The untold story behind our unspoiled views and virgin forests is this: these landscapes were inhabited, their features named, their forests utilized, their plants harvested and animals hunted. Native Americans have a history in our national parks measured in millennia. They were forcibly removed, and later treaty rights to traditional use such as hunting and fishing were erased, often without acknowledgment or compensation. Immediately after these removals, the parks were advertised as a showcase of uninhabited America, nature's handiwork unspoiled."

Additionally, the call for the removal of local indigenous populations from their traditional homelands in order to protect nature violates the rights of those being targeted for displacement. Moreover, the conception of wilderness devoid of human beings ignores the fact that for millennia, indigenous peoples actively modified local landscapes to suit their economic and political needs (Anderson, 2005; Chacon & Mendoza, 2012; Denevan, 1992; Mann, 2005; Merchant, 2007; Moss et al., 2001; Spence, 1996, 1999). As Kirch (2007: 86) notes, "[h]uman societies do not passively 'adapt' to their environments, they are actively engaged with those environments in a constant process of reciprocal feedback. As such, landscapes are socially constructed as well as physically modified." Consequently, attempts to understand ecosystem functions, and their management, should be informed by the fact that 'nature' is in effect made up of a "historicized, politicized, and humanized ecology" (Erickson

2006: 265). Thus, the Yellowstone Conservation Model violates the rights of the people who are displaced in order to "protect nature," but it also hampers our ability to truly understand complex human-environmental interactions (Loendorf & Stone, 2006; Nabokov & Loendorf, 1984).

Unfortunately, the misguided and unethical Yellowstone Conservation Model provided the archetype for later conservation efforts around the globe, and it continues to shape modern-day conceptions of wilderness (Merchant, 2003; Spence, 1999). For example, the US Congress passed the 1964 *Wilderness Act* which defined wilderness as areas "where man is a visitor *who does not remain*" (cited in Merchant 2003: 381, emphasis added).[8] The following example illustrates how some scientists have embraced this embraced this false dichotomy separating human beings from nature: While conducting fieldwork in and around Peru's Manu National Park, anthropologist Michael Alvard noted that a biologist, who was studying the nesting behavior of local birds, became upset with a band of Yora Indians for having "disturbed" his study. According to Alvard (1997: 610), "[t]he researcher was angered that the project was ruined when the Yora collected the bird eggs on the study site. Implicit in the anger was the view that humans were not part of the natural ecology of the area. Since humans have probably inhabited the area for several thousand years, human predation has potentially been an important selective force on these birds." Such an example illustrates vividly the need for scholars, development workers, and government personnel to reject the Yellowstone Conservation Model and to replace it with a broader perspective that includes native peoples as key ecosystem components.

Given the widespread adoption of the Yellowstone Conservation Model, which views native populations as environmental hazards, the significance and timeliness of *People, Parks, and Power: The Ethics of Conservation-related Resettlement* cannot be overstated. This publication exposes how marginalized indigenous peoples are currently being subjected to coercive forms of conservation (that were developed in the nineteenth century) to "protect nature."[9] This work also documents the arguments employed by elites to justify the dispossession of lands from contemporary indigenous populations.

It is important to keep in mind that marginalized populations of all racial and ethnic backgrounds can be harmed by the creation of exclusionary parks. For example, starting in 1825, emancipated African Americans living in New York City began purchasing land. Over time, black landowners formed a prosperous settlement known as Seneca Village. Additionally, in the 1840s, Irish immigrants fleeing the potato famine also began setting up residence at Seneca (Blakinger, 2016; Rosenzweig & Blackmar, 1992).[10] However, by the mid-1800s, city elites began calling for the creation of a landscaped area in New York that would be similar to the grand public parks found in many European capitals. "Most of the affluent merchants, bankers, and landowners who led the park campaign...desired a fashionable and safe public place where they and their families could mingle and promenade with other members of New York's upper crust, away from the crowded, dusty streets of downtown Manhattan" (Slavicek, 2009: 17). Consequently, the state legislature used eminent domain to confiscate more than 800 acres of land that would become Central Park. Thus, in 1855, Seneca Village residents were unceremoniously evicted from their homes (some violently). Prior to its destruction, the community had a population of 264 inhabitants (with 30% being Irish American), three churches, two schools, and three cemeteries (Martin, 1997).

Poor white farmers of the Blue Ridge Mountains of Virginia suffered a similar fate after elite conservationists decided that their Appalachian homeland would make a suitable playground for East Coast residents. As a means for justifying their removal, a sociological study described the isolated mountain communities of this region as being at low levels of "cultural development" (Sherman & Henry, 1933: 5) and attributed their poor living conditions to their isolation from modern American society (Sherman & Henry, 1933). The authors of the study wrote disparagingly of the region's isolated moutain people as indicated by the following: "'Social evolution' presumably still goes on but so slowly do groups [i.e. isolated Appalachian communities] go forward under their own power that no movement can be discerned through generations..." (Sherman & Henry, 1933: 9). Likewise, newspapers promulgated disparaging stories of mountain folk claiming that their way of life was backwards and unhealthy. In such a way, conservationists, social scientists, and the media created a narrative that promoted the notion that the only way to "save" such isolated Appalachian populations was to remove them from their mountain residences (Wild Song Cybil Productions, 2020; WVPT Living Virginia, 1998).

This collective effort proved fruitful when hundreds of Appalachian Mountain folk "were forced from their homes in the 1930s to make way for the Shenandoah National Park as state authorities used eminent domain to acquire private property that would be turned over to the federal government" (Lohmann, 2017: 1). Park advocates justified this action by describing the targeted region as being devoid of human settlements. However, given what has been reported above, this claim was not tenable. "In truth, the area was far from uninhabited. A 1934 census showed that approximately 435 families needed to be relocated before the park was dedicated in 1936, ..." (Lohmann, 2017: 8). Many mountain families were forced out with eviction notices along with "a visit from local law enforcement" (Lohmann, 2017: 8). Some displaced individuals never fully recovered emotionally over being banished from the home and mountains they loved (Lohmann, 2017).[11]

In stark contrast to the Fortress Conservation Model described above, Sapignoli and Hitchcock call for the establishment of mutually respectful dialogue and cooperation between conservationists and local stake-holding residents. In effect, they join a growing number of scholars who call for the decolonization of conservation (see Mbaria & Ogada, 2017).[12] This action will provide a foundation for instituting ethical, effective, and enduring conservation protocols.

Notes

1. Guha (1989: 71) argues that "the implementation of the wilderness agenda is causing serious deprivation in the Third World."

2. John Muir embraced Marsh's views as the famous conservationist felt "that preservation of nature and human occupation were incompatible" (Kantor, 2007: 46).
3. In fact, evidence of longstanding occupation of the Yellowstone area by Native Americans is provided by the considerable number of prehistoric archaeological sites recorded within the park (Kantor, 2007).
4. In 1868, the Shoshone Sheepeaters negotiated a treaty with the federal government ensuring the band's right to hunt in Yellowstone (Merchant, 2007).
5. These "Indian troubles" began in 1877 when 2000 soldiers pursued the Nez Perce through Yellowstone during the tribe's ill-fated attempt to escape to Canada. While traversing the park, warriors attacked and took several tourists hostage and killed two individuals (Kantor, 2007; NPS, 2019; Spence, 1996).
6. Surreptitious entry into Yellowstone by Native Americans for the purpose of harvesting natural resources continued into the late 1890s (Spence, 1999).
7. In 1868, when Muir encountered Native Americans in the Sierras, he wrote disparagingly of them stating that they were "mostly ugly, and some of them altogether hideous," having "no right place in the landscape" (Muir cited in Kantor, 2007: 46–47). He did go on to express admiration for several Indian guides on his 1879 and 1880 excursions to Alaska; however, he never rescinded any of his earlier derogatory statements regarding California Indians (Merchant, 2003).
8. Kantor (2007: 44) rightly points out that the 1961 *Wilderness Act* completely disregards the long history of indigenous occupation of many areas that were purportedly uninhabited. For example, see Doyle (2019) for the Crow's long-term habitation of the Absaroka-Beartooth Wilderness area that forms part of the Greater Yellowstone Ecosystem.
9. See Terborgh (1999) and Wilson (2014) who call for the removal and exclusion of local groups from protected areas.
10. According to Speed (2015), 50% of the heads of households in Seneca Village owned the land they lived on.
11. In 1998, WVPT Living Virginia produced a documentary titled "Displaced Mountain People" which provides a detailed account of this land grab committed by the US Government.
12. Soulé (1986) in his fine book on conservation biology calls for a balanced approach to conservation and protection of land and resources.

Winthrop University, Rock Hill, SC,   Richard J. Chacon
USA

References

Alvard, M. (1997). Comments. *Current Anthropology, 38*(4), 609–611.

Anderson, M. K. (2005). *Tending the wild: Native American knowledge and the management of California's natural resources*. University of California Press.

Blakinger, K. (2016). A look at Seneca village, the early Black settlement obliterated by the creation of Central Park. *New York Daily News,* May 17, 2016.

Borgerhoff, M., & Coppolillo, P. (2005). *Conservation: Linking ecology, economics and culture*. Princeton University Press.

Brockington, D. (2002). *Fortress conservation: The preservation of the Mkomazi Game Reserve, Tanzania*. James Currey.

Chacon, R., & Mendoza, R. (2012). *The ethics of anthropology and Amerindian research: Reporting on environmental degradation and warfare*. Springer.

DeGeorges, P., & Reilly, B. (2008). *A critical evaluation of conservation and development in Sub-Saharan Africa*. The Edwin Mellen Press.

Deguignet, M., Juffe-Bignoli, D., Harrison, J., MacSharry, B., Burgess, N., & Kingston, N. (2014). *United Nations list of protected areas*. UNEP-WCMC.

Denevan, W. (1992). The pristine myth: The landscape of the Americas in 1492. *Annals of the Association of American Geographers, 82,* 369–385.

Dowie, M. (2009). *Conservation refugees: The hundred-year conflict between global conservation and native peoples*. Massachussetts Institute of Technology.

Doyle, S. (2019). In home land. *Mountain Journal,* November 28, 2020. Accessed December 2, 2022, from https://mountainjournal.org/public-lands-were-home lands-for-the-crow-nation

Erickson, C. (2006). The domesticated landscapes of the Bolivian Amazon. In W. Balée & C. Erickson (Eds.), *Time and complexity in historical ecology* (pp. 235–278). Columbia University Press.

Guha, R. (1989). Radical American environmentalism and wilderness preservation: A third world critique. *Environmental Ethics, 11,* 71–83.

Igoe, J. (2004). *Conservation and globalization: A study of National Parks and indigenous communities form East Africa to South Dakota*. Thomson Wadsworth.

International Union for the Conservation of Nature and Natural Resources (IUCN). (2018). *Protected planet report 2018*. International Union for the Conservation of Nature and Natural Resources.

Juffe-Bignoli, D., Burgess, N., Bingham, H., Belle, E., de Lima, M., Deguignet, M., Bertzky, B., Milam, J. Martinez-Lopez, E., Lewis, A., Eassom, S., Wicander, J., Geldmann, A., van Soesbergen, A., Arnell, B., O'Connor, S., Park, Y., Shi, F., Danks, B., MacSharry, & Kingston, N. (2014). *Protected planet report 2014*. UNEP-WCMC.

Kantor, I. (2007). Ethnic cleansing and America's creation of National Parks. *Public Land and Resources Law Review, 28,* 42–64.

Kay, C. (1998). Are ecosystems structured from the top-down or bottom-up: A new look at an old debate. *Wildlife Society Bulletin, 26*(3):484–498.

Kirch, P. (2007). Three islands and an archipelago: Reciprocal interactions between humans and island ecosystems in Polynesia. *Earth and Environmental Science Transactions of the Royal Society of Edinburgh, 98*, 85–99.

Lohmann, B. (2017, March 20). Lost mountains. *The Roanoke Times, 1*, 8.

Loendorf, L. L., & Stone, N. M. (2006). *Mountain spirit: The sheep eater Indians of Yellowstone*. University of Utah Press.

Mann, C. (2005). *1491: New revelations of the Americas before Columbus*. Knopf.

Marsh, G. P. (2012) [1864]. *Man and nature: Or, physical geography as modified by human action*. General Books.

Martin, D. (1997, January 31). A village dies, a park is born. *New York Times*.

Mbaria, J., & Ogada, M. (2017). *The big conservation lie*. Lens & Pens Publishing.

Merchant, C. (2003). Shades of darkness: Race and environmental history. *Environmental History, 8*(3), 380–394).

Merchant, C. (2007). *American environmental history*. Columbia University Press.

Moss, K., Tippitt, A., Watts, S., May, A., & Simmons, S. (2001). *Journey to the Piedmont past source book*. Schiele Museum of Natural History.

Muir, J. (1916). *My first summer in the Sierra*. Houghton Mifflin Company.

Nabokov, P., & Loendorf, L. (2004). *Restoring a presence: American Indians and Yellowstone National Park*. University of Oklahoma Press.

National Park Service. (2019). Yellowstone: Flight of the Nez Perce. Accessed 30 June, 2021, from https://www.nps.gov/yell/learn/historyculture/flightnezperce.htm

Rosenzweig, R., & Blackmar, E. (1992). *The park and the people: A history of Central Park*. Cornell University Press.

Sahagun, L. (2014). John Muir's legacy questioned as centennial of his death nears. *Los Angeles Times*, Local/California Section. November 13, 2014.

Sapignoli, M., Hitchcock, R. K., & Mangiameli, G. (2017). Introduction. In M. Sapignoli, R. Hitchcock, & G. Mangiameli (Eds.), *Popoli Indigeni in Africa: Articolazioni Globali Locale e Nazionali* (pp. 1–22). UNICOPLI.

Sherman, M., & Henry, T. (1933). *Hollow folk*. Thomas Y. Crowell Company.

Slavicek, L. (2009). *New York city's Central Park*. Infobase Publishing.

Soulé. M. (1986). *Conservation biology: The science of scarcity and diversity*. Sinauer.

Speed, B. (2015). New York destroyed a village full of African-American landowners to create Central Park. *City Metric*. March 30, 2015. Accessed (8/5/2017).

Spence, M. (1996). Dispossessing the wilderness: Yosemite Indians and the National Park Ideal, 1864-1930. *Pacific Historical Review, 63*(1), 27–59.

Spence, M. (1999). *Dispossessing the wilderness: Indian removal and the making of the National Parks*. Oxford University Press.

Terborgh, J. (1999). *Requiem for nature*. Island Press.

Treuer, D. (2021). Return the National Parks to the tribes. *The Atlantic,* May 20, 2021.

Wild Song Cybil Productions. (2020). Shadows. Wild Song Cybil Productions.

Wilson, E. O. (2014). *A window on eternity: A biologist's walk through Gorongosa National Park.* Simon & Schuster.

WVPT Living Virginia. (1998). *Displaced mountain people. Shenandoah Valley Education Television.* Accessed September 13, 2022, from https://www.youtube.com/watch?v=NkjWunrZD7I

Preface

Bau, a Ju/'hoan San woman, stirred as she awoke an hour before sunrise next to a small smoking fire in the Tsodilo Hills of northwestern Botswana in late 1994. She got up to tend the fire, and looked at her husband, C'unte Gcau, and their two small children, a boy, 5, and a girl, 3, sleeping on skin mats next to the fire where she had been herself a few minutes before. She walked over to a small grass hut about 20 ft away and collected some twigs that were in a pile outside the door and laid them carefully on the fire with a bit of grass. Once the fire was going, she placed a handful of mongongo nuts (*Schinziophyton rautanenii*) in the embers to cook.[1]

Fire (*dà'á* in Ju/'hoan) is created by rubbing two sticks together and lighting a tuft of grass, which a man or woman blows on and then places the burning grass in a hearth with some grass and small sticks to get a fire going. In some cases, a fire was started by striking a flint with a small piece of metal, creating sparks that drop into tufts of grass.

The camp in which she was living consisted of 7 families, and a total of 40 people. As she warmed herself by the fire, her husband, C'unte, got up and walked to the side of the cleared area and relieved himself. He then walked back to help Bau with the cooking. The two children stayed asleep until after sunrise.

Bau and C'unte were joined by two other women and a man who then talked about their plans for later in the morning. Bau and her friends wanted to go into the bush to gather nuts, fruits, and roots. C'unte said he and two other men would collect their belongings from the nearby huts (skin bags, quivers, bows, poisoned arrows, spears, and sticks for making fires) and would set out to look for wild animals. They planned to walk several kilometers to where a Ju/'hoan friend of theirs had seen the tracks of three hartebeest (*Alcelaphys buselaphus*), which they would then follow until they got within sight of them.

The hunters would sneak up through the low bushes and grass until they were within arrow shot, a distance of some 25 m. When the hartebeest were grazing, they would stand up and let their small arrows fly. If they were successful in hitting one, they would follow the animal until it tired from the chase and the effects of the poison and then finish it off with spears. They would butcher the animal, which would take at least two hours. They would cook some of the innards such as the heart and liver over a small fire and eat them while they were processing the animal. They would hang some of the meat on tree branches to dry, making what is known in southern Africa as biltong (dried meat). Once they had finished, they would collect their belongings and the meat and take everything back to their camp below what is known as the Male Hill of Tsodilo (18°45'40" S, 21°44'45" E)

In the meantime, the three women would be in the mongongo groves on the dunes collecting mongongo nuts that had fallen from the trees and putting them in leather bags (karosses) that they had strapped over their shoulders. The women would also be looking for edible plants and roots which they would dig up using digging sticks and their hands. If they were thirsty, they would take a sip of water from the ostrich eggshells that they were carrying. When they ran out of water, they would go to a marula (*Sclerocarya caffra*) or mongongo tree and, using straws made of *Grewia*, sip water out of the holes in the trunks. Ju/'hoan men tended to go farther than women in their quest for wild animals and plant foods, a pattern similar to that of the Hadza in the Lake Eyasi region of Tanzania (see Wood et al., 2021).

After several hours of collecting wild plants and the occasional small animal such as a tortoise, the women would return to their camp in Tsodilo and lay the fruits of their labors out on skins on the ground to allow them to dry. Some of the nuts, fruits, and seeds would be placed in a small wooden mortar and pounded using a small pestle, also made of wood. They would mix the pounded plant meal with a bit of meat from animals that their husbands and friends had brought home and cook the mixture in water in a small iron pot on the fire. Children would come to the fire and try and sample the boiling food, but their parents would shoo them away. Ju/'hoan adults tried to make sure that the children did not get too close to the fire or the iron pot in order to avoid accidents. Some adults and teenagers had scars from burns that they had suffered when they were younger. Once the food was ready, the women would put it in small wooden or metal bowls and share it with the people who had assembled near the fire.

Conversations around the fire covered a wide variety of topics.

One woman, /Tikay, talked about an experience she had when out gathering the previous day when she and her friends came across the tracks of two lions (*Panthera leo*). She said that the women returned to camp immediately to tell the others about the presence of the predators. An older man, Shoroka Cuntae, suggested that he and some of his male friends should go out and see where the lions were and see whether they had killed anything. If they had, he recommended that the men wait until the lions were full from their meals. He then said that they should run up to the lions and scare them off their kill, taking the leftover meat for their own use. He said that the men would have to keep a watchful eye on the lions in case they decided to return to their prey.

Another woman talked about seeing tracks of people whose footprints she did not recognize. She said that she told the *n!ore kxau* (an elderly person in the camp who was the "manager" of the land and its resources) about the tracks. The community would then have to decide about whether to go out and see who the visitors to their territory, known locally as a *n!ore*, were and have a talk with them and on that basis decide if they should be allowed to stay in the area with them and use the local resources.

One of the biggest concerns that they had was that the Botswana government was pushing for the land of places such as Tsodilo to be declared off-limits to local residents, opening them up exclusively for tourists and private companies. The Ju/'hoansi were aware that they and their neighbors, the Mbukushu, had been affected by the establishment of the Tsodilo Hills National Monument. They knew of plans to declare the Tsodilo Hills either a biosphere reserve or a World Heritage Site.[2] Several Ju/'hoansi said that they were worried about what they had heard happened to the G/ui/ G//ana, Bakgalagadi, and other groups in the Central Kalahari Game Reserve, also in Botswana, with government efforts to relocate people out of the protected area.

The talks around the fire were lively. Sometimes all of the adults spoke at once, telling stories and joking with one another. The stories that they related, some of which were about past experiences, conveyed information that was useful for people in the group to know. A number of the stories they told were especially for children, some of whom sat next to their parents around the fire. There were tales about spirits and the abilities of some Ju/'hoansi to heal others through going into trance or engaging in other kinds of rituals. Some of the stories dealt directly with issues of land and resource rights and management.

Stories were told about places on the landscape such as water holes (pans) that might contain water and would provide important sources of drinking water for people as they moved from one place to another. This was the dry season at Tsodilo, and places that contained water were few and far between. The men talked about planning a hunt in places to the south of the hills, where wildebeest (*Connochaetus taurinus*), gemsbok (*Oryx gazella*), kudu (*Tragelaphus strepsiceros*), and buffalo (*Syncerus caffer*) had been spotted coming to a pan to drink.

Some of the Ju/'hoansi wanted to go over 150 km to the south to a pan called ‡Gi that was located close to !Ubi on the Botswana–Namibia border that contained water year-round where animals were known to come, especially in the winter dry season (Brooks, 1978; Helgren & Brooks, 1983). At ‡Gi, Ju/'hoansi men had constructed small hides, known as blinds or hunting stands, which were made of calcrete blocks, rocks, twigs from local trees, and grasses. The blinds were places where two to three hunters could hide at night, waiting for animals to come down to drink. This kind of ambush hunting technique was used primarily in the dry season, when there were fewer places in the bush where animals could find water (for a discussion of these hunting blinds and their use, see Hitchcock et al., 2019).

While hunting, which was done mostly by men, was an important source of food for the Ju/'hoansi, the bulk of the food supply among the Ju/'hoansi consisted mainly of plants, which were largely provided by women. Ju/'hoan children, they said, generally did not contribute very much to the groups' daily diet, spending much of their time playing in the camps. Sometimes older children would go with their mothers and other women and sometimes men on gathering trips. Back in camp, the older children would take care of their younger siblings. Infants were carried on the backs of their mothers when they traveled or went on gathering expeditions. The backpack of a woman consisted of a skin or cloth wrap to hold the baby, and a skin carrying bag which held a digging stick, fire-making sticks, sinew for wrapping around grass or plants they found, and often a tortoise shell powder puff for use for cosmetic purposes.

The traditional land management system of the Ju/'hoansi San (the *n!ore* system) is structured in such a way that individuals and families have rights to territories (*n! oresi*) which they utilize for purposes of obtaining food, fuel, building materials, medicines, and other goods. The rights to these territories are handed down from one generation to the next, and people are often able to obtain permission to use other people's territories if they seek those rights from the local elders and land managers (*n!ore kxausi*). Those "resource managers" as they have sometimes been described would poll their camp-mates to see what their opinion was. If it was decided that the resource densities in their areas were too low, they would opt not to give permission to the people making the request. In most cases, however, the decision was made to allow other people to come into the *n!ore* to share the resources. In the case of the Ju/ 'hoansi in the Tsodilo Hills, the average territory size was, according to the Ju/ 'hoansi, about 225 km^2 in size. In drought years, the territories would expand to 400–450 km^2 in area, with people moving toward the Okavango Delta to the east or to Nxau Nxau, a Ju/'hoan community to the southwest, and back again.

This flexibility in land use systems helps to ensure that people do not starve, and it serves to reinforce social alliances among groups and individuals. From the perspective of the Ju/'hoansi, the "true people" they were careful resource managers who attempted to avoid over-exploiting their resources and ensuring, if at all possible, the long-term availability of the plant, animal, and mineral resources that sustained them.

Little did this small group of Ju/'hoansi realize that in May 1995 they would be relocated away from the Tsodilo Hills by the Botswana government and paid a small amount of compensation, approximately P17,500 at that time (equivalent to US$6340.58 in May 1995) for the group of 40 Ju/'hoansi. They were given a borehole for domestic water and for their livestock in an area located approximately 5 km south of the Male Hill. Not long afterward, the borehole caved in, and they had to walk 10 km round-trip to obtain water from the borehole at the National Museum headquarters in the Tsodilo Hills and carry the water back to the Ju/'hoan village in buckets on their shoulders. Their incomes from craft sales declined by 90%, in part because they were no longer close to where the tourists entered the hills and most of the funds were captured by the group of 60 Mbukushu allowed to live close to the southern gate to the hills. Individual Ju/'hoansi who wanted to give tours of the rock art had to walk 5 km to the Mbukushu village and wait next to the gate to the World Heritage Site to see if tourists visiting the hills would hire them.

Ju/'hoan cattle, goat, and donkey numbers declined because of lack of sufficient water and poorer grazing south of the hills. All of these events occurred between 1995, when the Ju/'hoansi were moved, and 1999. During a visit to the Hills in 1999, one of us (Hitchcock) learned that the borehole had been fixed, and people did not have to go to the National Museum campground to get water. The Ju/'hoansi

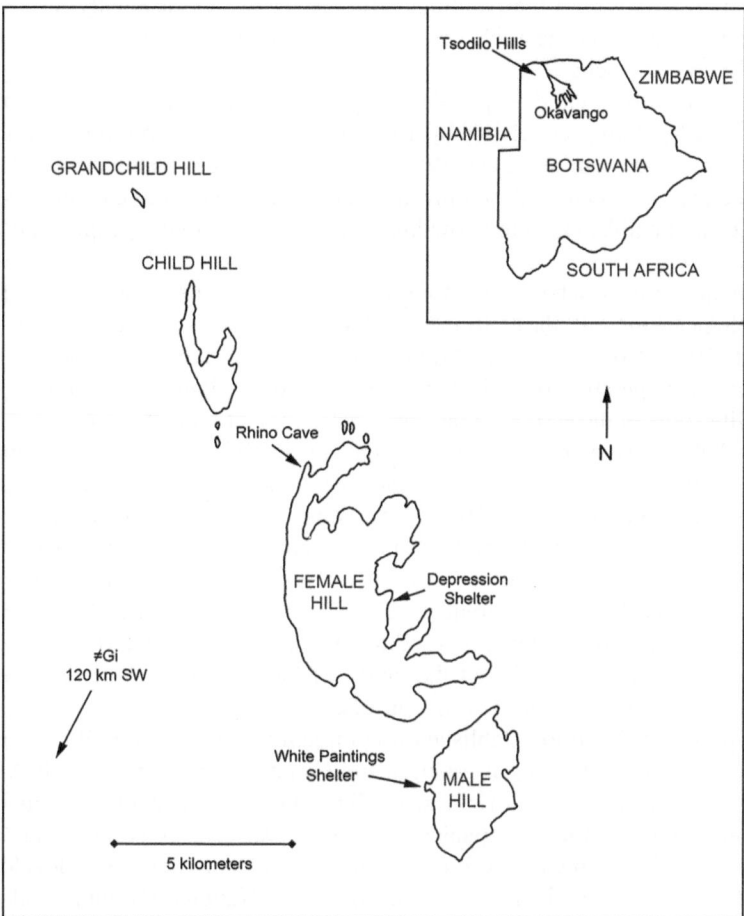

Fig. 1 Tsodilo Hills World Heritage Site in Botswana

maintained at the time that while they used to be the main guides for tourists, much of those responsibilities had been taken over by their former neighbors, the Mbukushu.

The Tsodilo Hills contain thousands of rock paintings at some 450 different sites which local Ju/'hoan and Mbukushu communities have various theories and conjectures about regarding their origins and significance (Campbell et al., 2010). There are some key archaeological sites where important finds were made, including White Paintings Shelter, Depression Shelter, and Rhino Cave (see Fig. 1 for the location of these sites). All of these sites were occupied by hunter-gatherers, some of them by the ancestors of contemporary people living in the Tosdilo Hills. According to interviews, there are claims that the paintings were made by God; others say that the San who lived there previously (the Ncaekhwe) painted them, and still others are uncertain about who did the paintings (Biesele, 1974). Occasionally, it is claimed that the red marks on the rock are from birds having consumed red berries and defacating on the rocks.

The Tsodilo Hills and their people and rock art came to international attention in the 1950s, when they were visited by a South African writer and filmmaker, Laurens Van Der Post (see Van Der Post, 1958; Van Der Post & Coulson, 1988). Van Der Post described the beauty of the hills and remarked on what he felt was a "spiritual presence" in the hills. He noted how important they were to the people who resided there.

There are serious debates about the meaning, importance, and dates of specific archaeological finds in places in the Tsodilo Hills (see, e.g., Robbins et al., 2007; Staurset, 2008; Coulson et al., 2011). There is no question, however, about the archaeological, geomorphological, and ethnographic significance of the Tsodilo Hills, with some of the earliest evidence of fish exploitation in the Kalahari (Robbins et al., 1994), early evidence of mongongo nut exploitation and use (Robbins & Campbell, 1990), some of the earliest evidence of the use of bone projectile points in the Kalahari (Robbins et al., 2011), and some of the most intensive mining of specularite (used for hair decoration) in southern Africa (Robbins, 2016). All of these sites are seen as important when it comes to cultural heritage and tourism (Keitumetse et al., 2007; Giraudo, 2011, 2016, 2017, 2018; Mbaiwa, 2016; Wendorff, 2022). Local people in Tsodilo have asked why Tsodilo was designated as a World Heritage Site without the inputs of local residents. The respone of the government has been that they were consulted.

The people of Tsodilo established a community trust, the Tsodilo Community Development Trust (TCDT) in 2006, in the Ngamiland (NG 6) Wildlife Manage-ment Area, which is 225 km^2 in size. They have also benefitted from mineral agreements with mining companies, including DeBeers Botswana and Tsodilo Resources, both of which have provided them with funds for water development, campsite establishment, roads, and a craft shop at the National Museum headquarters site which has helped some Tsodilo residents to increase their incomes and provided outlets for their craft products.

The numbers of tourists visiting the hills declined precipitously after March 2020 when the Government of Botswana stopped international travelers from visiting the country as a result of the coronavirus pandemic. The coronavirus pandemic did not, however, stop the operations of oil companies, who were carrying out prospecting activities in the area north of the Tsodilo Hills. ReconAfrica, a Canadian oil and gas company, announced plans to drill for oil in the area north of Tsodilo in September 2020 (Barbee with Nash, 2020; Barbee & Nemee, 2020; York & Gravey, 2021; ReconAfrica website, accessed 30 June 2022). In July 2022, people in the Tsodilo Hills were informed by company spokespersons that they would likely be relocated to the south as a result of the planned exploratory operations. Local residents, including members of the Tsodilo Community Development Trust and the Tsodilo village development committee, said that ReconAfrica personnel had been to see them again in December 2022 and that they were told in no uncertain terms that they faced relocation due to the drilling operations.

Representatives of indigenous support organizations in Shakawe and Maun and international environmental organizations have already called attention to the potential negative consequences the drilling operations may have on local water systems, wildlife, birds, and people (York & Greney, 2021; Gakemotho Wallican Satau, personal communication, December 22, 2022). The oil explorations in Botswana had yet to start as of April 2023, but discussions continue at the district and national levels. Questions about the viability of the ReconAfrica operations in Botswana continue to be raised (Goodell, 2023).

From the standpoint of the contemporary Ju/'hoan and Mbukudu residents of the Tsodilo Hills, the area is not only rich in natural and cultural resources but is also a significant place on the landscape which to them is sacred. Tsodilo has shaped their past and present and holds great promise for the future. The question that remains, however, is who has the power to control decision-making involving Tsodilo: local community members, the national government of Botswana, or outside interests including private mining and tourism companies?

Notes
1. The mongongo underwent a name change, from *Ricinodendron rautanenii* to *Schinziophyton rautanenii* (Schinz) two decades ago.
2. The Tsodilo Hills were in fact declared a World Heritage Site by the UNESCO World Heritage Commission in 2001, setting aside an area of 48 km^2 in size, with a buffer zone of 704 km^2 (see Basinyi, 2019).
3. For a discussion of what transpired in the cases relating to the Central Kalahari Game Reserve, see Sapignoli (2018) and below.

Milan, Italy   Maria Sapignoli
Albuquerque, NM   Robert K. Hitchcock

References

Barbee, J., & Nash, J. (2020). Mystery surrounds plans to start fracking near Namibia's Kavango River and Botswana's Tsodilo Hills. *National Geographic Society*, 16 September 2020.

Barbee, J., & Neme, L. (2020). Oil drilling, possible fracking planned for Okavango Oil drilling, possible fracking planned for Okavango region—elephants' last stronghold. *National Geographic Society,* 28 October 2020.

Basinyi, S. (2019). *Living with heritage: The case of Tsodilo World Heritage Site and neighbouring localities*. Archaeopress Publishing Ltd.

Biesele, M. (1974). A note on the beliefs of modern bushmen concerning the Tsodilo Hills (Paintings). *Newsletter of the South West Africa Scientific Society, 15*, 3/4 (1–3).

Brooks, A. (1978). A note on late stone age features at /Gi: Analogies from Historic San hunting practices. *Botswana Notes and Records, 10*, 1–3.

Campbell, A., Robbins, L., & Taylor, M. (Eds.). (2010). *The Tsodilo Hills: Copper bracelet of the Kalahari*. Michigan State University Press and Botswana Society.

Coulson, S., Staurset, S., & Walker, N. (2011). Ritualized behaviour in the Middle Stone Age: Evidence from the Tsodilo Hills, Botswana. *Paleoanthropology, 2011*, 18–61.

Goodell, J. (2023). *The neocolonial oil racket*. National Geographic Society, 26 March 2023.

Giraudo, R. F. (2011). *Intangible heritage and tourism development at the Tsodilo Hills world heritage site*. Ph.D. dissertation, University of California, Berkeley, California.

Giraudo, R. F. (2016). World heritage, tourism development, and identity politics at the Tsodilo Hills. In L. Buurdeau, M. Gravari-Barbas, & M. Robinson (Eds.), *World heritage sites and tourism: Global and Local Relations*, (pp. 77–93). Routledge.

Giraudo, R. F. (2017). Heteroglossic heritage and the first-place of the Kalahari. *International Journal of Heritage Studies, 24*(2), 128–141.

Giraudo, R. F. (2018). Performing our past to secure our future: A look at san-owned cultural tourism in the Kalahari. In R. F. Puckett & K. Ikeya (Eds.), *Research and activism among the Kalahari San today: Ideals, challenges, and debates* (pp. 215–230). Senri Ethnological Studies 99. National Museum of Ethnology.

Helgren, D. M., & Brooks, A. S. (1983). Geoarchaeology at ǂGi: A Middle Stone Age and Later Stone Age site in the North-west Kalahari. *Journal of Archaeological Science, 10*, 181–197.

Hitchcock, R. K., Crowell, A. L., Brooks, A. S., Yellen, J. E., Ebert, J. I., & Osborn, A. J. (2019). The ethnoarchaeology of Ambush Hunting: A case study of ǂGi Pan, Western Ngamiland, Botswana. *African Archaeological Review, 36*, 119–144.

Keitumetse, S. O., Matlapeng, G., & Monamo, L. (2007). Landscape, communities and world heritage: Pursuit of the loçal in the Tsodilo Hills, Botswana. In D. Hicks, L. Mcatackney & G. Fairclough (Eds.), *Envisioning landscape: Situations and standpoints in archaeology and heritage* (pp. 101–119). Walnut Creek: LeftCoast Press.

Mbaiwa, J. (2016). The commodification of world heritage sites: The case study of Tsodilo Hills in Botswana. In M. Haratsebe, N. Moswete, & J. Saarinen (Eds.), *Cultural tourism in Southern Africa* (pp. 101–120). Channel View Publications.

Robbins, L. H. (2016). Sebilo: 19th century hairdos and ancient specularite mining in Southern Africa. *International Journal of African Historical Studies, 4*(1), 103–131.

Robbins, L. H., Brook, G. A., Murphy, M. L., Campbell, A. C., Melear, N., & Downey, W. S. (2000a). Late quaternary archaeological and palaeoenvironmental data from sediments at Rhino Cave, Tsodilo Hills, Botswana. *Southern African Field Archaeology, 9*, 17–31.

Robbins, L. H., & Campbell, A. C. (1990). Prehistory of Mongongo nut exploitation in the Western Kalahari Desert, Botswana. *Botswana Notes and Records, 22*, 37-42.

Robbins, L. H., Campbell, A. C., Book, G. A., & Murphy, M. L. (2007). World's oldest ritual site? The 'Python Caves' at Tsodilo Hills world heritage site, Botswana. *Nyame Akuma, 67,* 2–6.

Robbins, L. H., Campbell, A. C., Brook, G. A., Murphy, M. L., & Hitchcock, R. K. (2012). The antiquity of the bow and arrow in the Kalahari Desert: Bone points from White Paintings Rock Shelter, Botswana. *Journal of African Archaeology, 10*(1), 7–20.

Robbins, L. H., Murphy, M. L., Stewart, K. M., Campbell, A. C., & Brook, G. A. (1994). Barbed bone points, paleoenvironment, and the antiquity of fish exploitation in the Kalahari Desert, Botswana. *Journal of Field Archaeology, 21,* 257–264.

Sapignoli, M. (2018). *Hunting justice: Displacement, law, and activism in the Kalahari.* Law and Society Series. Cambridge University Press.

Staurset, S. (2008). *Becoming human: Ritualized behaviour and Middle Stone Age points – A case study from Rhino Cave, Tsodilo Hills, Botswana.* Masters thesis, University of Oslo, Oslo, Norway.

Taylor, M. (2010). The politics of cohabitation: Social history of Tsodilo. In A. Campbell, L. Robbins, & M. Taylor, (Eds.), *The Tsodilo Hills: Copper bracelet of the Kalahari* (pp. 116–125). Michigan State University Press and Botswana Society.

Taylor, M. (2014). 'We are not taken as people': Ignoring the indigenous identities and history of Tsodilo Hills World Heritage Site, Botswana. In S. Disko & H. Tugendhat (Eds.), *World heritage sites and indigenous peoples' rights* (pp. 119–129). International Work Group for Indigenous Affairs.

Van Der Post, L. (1958). *The lost world of the Kalahari.* Morrow and Company.

Van Der Post, L., & Coulson, D. (1988). *The lost world of the Kalahari.* Chatto and Windus.

Wendorff, M. (2022). The Tsodilo Hills: A multifaceted world heritage site. In F. D. Eckardt (Ed.), *Landscapes and landforms in Botswana* (pp. 345–360). Springer.

Wiessner, P. W. (2014). Embers of Society: Firelight talk among the Ju/'hoansi Bushmen. *Proceedings of the National Academy of Sciences, 111*(39), 14027–14035.

Wood, B. M., Harris, J. A., Raichlen, D. A., Pontzer, H., Sayre, K., Sancilio, A., Berbesque, C., Crittenden, A. N., Mabulla, A., McElreath, R., Cashdan, E., & Jones, J. H. (2021). Gendered movement ecology and landscape use in Hadza hunter-gatherers. *Nature Human Behavior, 5,* 436–446.

York, G., & Graney, E. (2021). As Calgary's Recon Africa drills for Namibian oil, a global outcry over endangered elephants grows. *Mail and Guardian,* 29 May 2021.

Acknowledgments

Support for some of the research upon which this book is based was provided by the Max Planck Institute of Social Anthropology, the University of New Mexico, the University of Milano, the US National Science Foundation (grant no. BCS 1122932), Hivos (The Netherlands), the International Work Group for Indigenous Affairs (IWGIA), the Open Society Initiative for Southern Africa (OSISA), the Nyae Nyae Development Foundation of Namibia (NNDFN), the Kalahari Peoples Fund, and Brot für die Welt (Germany).

We want to thank Rick Chacon for his unwavering support of our work, as well as Sujatha Chakkala, Shinjini Chatterjee, and other editors of Springer Books for their assistance. We wish to express our deepest appreciation to the people of southern Africa and the governments of Botswana, Namibia, and Zimbabwe for permission to conduct some of the work on which this book is based. The figures for this book were drawn by Marieka Brouwer-Burg of the University of Vermont. Editorial suggestions for improvement were provided by Rick Chacon and Melinda C. Kelly.

We also wish to thank the various organizations who have provided useful information, including the International Work Group for Indigenous Affairs, the Forest Peoples Programme, Cultural Survival, Survival International, the Minority Rights Group International, Conservation International, the World Wildlife Fund (US), the Worldwide Fund for Nature (Switzerland), the Nature Conservancy, the Norwegian Agency for Development Coopartion (NORAD), the Food and Agriculture Organization (FAO), the United Nations Development Programme (UNDP), the United Nations High Commissioner for Refugees (UNHCR), the United Nations Environment Programme (UNEP), the United States Agency for International Development (USAID), and the World Bank. We also wish to extend our appreciation to the various indigenous peoples who provided us with information, insights, and ideas.

And last but by no means least, we would like to thank the various indigenous people, government and NGO personnel,ecologists, conservationists, and social scientits with whom we have worked over the past 16 years.

This book is dedicated to Sir David Attenborough and to the memories of Giorgio Sapignoli and John and Karleen Hitchcock.

Contents

About the Authors

Maria Sapignoli is Assistant Professor in Cultural and Social Anthropology in the Department of Philosophy, Piero Martinetti, University of Milan in Italy. She holds a BA and MA in Anthropology from the University of Bologna in Italy and a PhD in Sociology from the University of Essex in the UK. Before joining the Department of Philosophy Piero Martinetti (University of Milan) in April 2021, she spent eight years at the Max Planck Institute for Social Anthropology in Germany, first as research fellow in its Law & Anthropology Department and most recently heading a Max Planck Independent Research Group. She is continuing to cooperate with the MPI as Cooperation Partner and as Accompanying Scientific Committee member of the research cluster she contributed to setting up, titled *Anthropology of AI in Policing and Justice*. She has done research in legal activism, indigenous rights, and social movements. Some of her work has been in international organizations, including the United Nations General Secretariat Headquarters, the United Nations Permanent Forum on Indigenous Issues, and the International Fund for Agricultural Development. She has also worked for several non-government organizations.

Over the last few years, she has been fellow in residence in several universities including McGill University (Canada), University of Milan (Italy), and New York University (USA). She has worked on issues of San activism in the Kalahari Desert of Southern Africa for 15 years. Dr Sapignoli is the author of a monograph, Hunting Justice: Displacement, Law, and Activism in the Kalahari (Cambridge University Press, 2018). She is co-editor of Palaces of Hope: The Anthropology of Global Organizations (Cambridge University Press, 2017), of *La Questione Indigena in Africa* (Unicopli, 2017), and of the *Oxford Handbook of Law and Anthropology* (Oxford University Press, 2022). In addition, she has published numerous articles and book chapters.

Robert K. Hitchcock is a professor in the Department of Anthropology at the University of New Mexico. He is also a board member of the Kalahari Peoples Fund (KPF), a non-profit 501©3 organization that provides funding for education, development, and capacity-building training for indigenous and minority peoples in southern Africa. He has a BA in Anthropology and History from the University of California, Santa Barbara, and an MA and PhD in Anthropology from the University of New Mexico. He has taught anthropology at several universities, including South Dakota State University, the University of Nebraska-Lincoln, Michigan State University, Truman State University, and the University of New Mexico. Originally trained as an archaeologist at UC Santa Barbara and the University of New Mexico, Hitchcock today is an applied cultural anthropologist, human ecologist, and socioeconomic development specialist. He has spent much of his professional career working on issues facing current and former hunters and gatherers and agropastoral peoples, particularly in southern and eastern Africa. A significant portion of his ethnographic and human rights work has been with the San (Bushmen) of southern Africa, especially those of Botswana, Namibia, and Zimbabwe, with whom he has worked for nearly five decades. Currently, he is serving as a social and environmental safeguards specialist for a United Nations Development Programme project devoted to managing the human–wildlife interface in the Kgalagadi and Ghanzi drylands of Botswana. Some of his work has been on human and indigenous peoples' rights and well-being, especially in protected areas and adjacent buffer zones.

Dr. Hitchcock is the author of *Kalahari Communities: Bushmen and the Politics of the Environment in Southern Africa* (IWGIA, 1996), co-author of *The Ju/'hoan San of Nyae Nyae and Namibian Independence: Development, Democracy, and Indigenous Voices in Southern Africa* (Berghahn, 2013), and a co-editor of *Information and Its Role in Hunter-Gatherer Bands* (Cotsen Institute of Archaeological Press, 2011), *Hunter-Gatherers and Their Neighbors in Asia, Africa, and South America* (Senri Ethnological Studies, 2016), and *Archaeology at the Threshold: Studies in the Processes of Change* (Florida University Press, 2023).

List of Figures

List of Tables

Chapter 1
Biodiversity Conservation, Protected Areas, and Indigenous Peoples

Introduction

On January 29, 2002, 29 heavy trucks and seven smaller Land Cruisers and Land Rovers drove along the dirt track to Mothomelo in the heart of the Central Kalahari Game Reserve (CKGR) in Botswana. This settlement at that time had only functioning borehole that provided water to the residents of the reserve. The government convoy had entered the CKGR in order to bring this service to an abrupt end, and at the same time to take the residents away from their old lives and move them to new places that had been prepared for them, beyond the gates of the reserve, where they could become 'modern' as the government put it.

To get to its destination, the convoy drove along what one observer called "one of the most kidney-jarring, axle-snapping, sand blasted and sun-burned landscapes on Earth" (Workman, 2009: 1). Where the track divided, the drivers had to pick a careful route around patches of deep sand that, especially for the smaller vehicles, could result in getting bogged down, resulting in difficult efforts to free them, and delay. Occasionally the convoy would stop for maintenance. Tall grass that accumulated in the undercarriages of vehicles with lower clearance had to be removed periodically before it caught fire from the heat of the engines.

Under these conditions, there had to be a resolute purpose behind the journey. On this occasion, the basic motivation was the removal of the San and Bakgalagadi from what the government explicitly viewed as a backward way of life. The details surrounding the forced removals of residents from the CKGR in 2002 were (and continue to be) contested, with some Botswana government officials maintaining through the years that the removals were consensual, orderly, and the outcome of intensive consultation with everyone concerned. Government officials claimed that tables had been set up at the edge of the villages and residents lined up to provide their consent to the relocation with the particulars of their names and places of birth and a thumb print on the documents that made their decision official. According to them, it was all very peaceful and "by the book," and compensation for the

© The Author(s) 2023
M. Sapignoli, R. K. Hitchcock, *Anthropology and Ethics*,
https://doi.org/10.1007/978-3-031-39268-9_1

inhabitants' losses was paid. Together with assertions of order and consent in the process of relocation, the government promoted a narrative of prosperity and development as a justification for the removals.

In fact, the removals were anything but orderly. When the government trucks arrived in the villages, the officials began destroying people's homes and corrals and proceeded to cut down their trees and rip up their gardens. Storage tanks and buckets containing water were overturned, and people's ostrich eggs, which they used as canteens for water, were crushed. Goats and dogs were chased down by the government workers and captured, tied up, and thrown on the waiting trucks.

Some village residents attempted to reason with the government officials in charge of the operation, but they were rebuffed, and the government workers continued to load people and animals onto the trucks. In some cases, people who attempted to resist were struck or warned that they would be beaten if they did not immediately get onto the trucks. In several cases people were struck if they refused to get onto the vehicles. Some people had children and other relatives who were thrown onto the trucks which later set out for different places, leading to separation of children from their parents in some cases. Government officials destroyed the village water pump, took the lifting rods and casings out of the borehole, and poured cement down the hole, covering it over with cement and they finally capped it with an iron plate. It took several months for people who had been removed from the Central Kalahari to trace down their relatives who were relocated to different places around the reserve.

Biodiversity Conservation and Involuntary Relocation

The Botswana government argued that the removals of the residents of the Central Kalahari were done in order to promote biodiversity conservation (Zips-Mairitsch, 2013; Sapignoli, 2018; High Court Transcripts, Roy Sesana case 2004–2006). It was suggested by the government lawyers that the presence of resident people and their animals was a threat to biodiversity conservation. Ecologists who had worked in the Central Kalahari said that the residents of the game reserve were over-exploiting the wild animal resources, with little, if any, evidence to support their claims (Owens & Owens, 1981, 1984). Subsequent research revealed that the claims about over-hunting were not accurate.

As Edward O. Wilson (1988: 3) noted 'Biological diversity must be treated more seriously as a global resources, to be indexed, used, and above all, preserved." Conservation of biodiversity has been seen as a priority by government entities, travelers, and individuals for over two centuries. One preservation strategy has been to set aside areas as protected, the first and perhaps most common biodiversity strategy Biodiversity refers broadly to the full set of species, variation within species, and the variety of ecosystems that contain species (Borgerhoff-Mulder & Coppollilo, 2005: 2). Other strategies include declaring specific species as off-limits to human utilization, two examples being elephants and whales. Wilson (2016) has recommended that half of the world should be set aside for conservation purposes (cf. Kopnina, 2016). The International Union for the Conservation of Nature and Natural Resources (IUCN) has suggested that 30% of all countries should have land designated as conservation land.

The forcible relocation of people from places designated as national parks and other kinds of protected areas is a process that has become all too common in Africa, Asia, North America and other parts of the world (see Dowie, 2005, 2009; Igoe & Brockington, 2006; Naughton-Treves et al., 2005; West et al., 2006; Tauli-Corpuz et al., 2018, 2020). In a number of cases, this relocation process has been accompanied by violence (Fanari, 2019; Mushonga & Matose, 2020; Ramutsindela et al., 2022). The people in those areas set aside as protected places have sometimes moved willingly, but there are also cases where local people have resisted the efforts of the state to relocate them. They have resorted to such actions as sit-down strikes, filing of legal cases, negotiations, and disappearing into the bush to evade capture and removal from their ancestral land.

For the purposes of this volume, we focus on those people who see themselves or are seen by others as indigenous. There is no international or academic consensus on the meaning of the terms 'indigenous,' 'indigenism,' or 'indigeneity' (Kuper, 2003; Niezen, 2003; Anaya, 2009; Merlan 2009). There is considerable variation on the terminology that is used to designate indigenous peoples. The International Labor Organization and Survival International use the term "tribal and indigenous peoples" (and in the past also used "semi-tribal peoples") while the World Bank and the United Nations prefer 'indigenous peoples.' Academics sometimes refer to them as native peoples, first peoples, marginalized peoples and aboriginal peoples. A significant number of governments in Africa and Asia do not recognize peoples within their borders as indigenous. Virtually all African states argue that all citizens of the country are indigenous, and this is true for nearly all countries in Asia. Central and South American countries, on the other hand, do tend to recognize some of the residents as indigenous (Brysk, 2000; Hall & Patrinos, 2012a).

The Independent Commission on International Humanitarian Issues (1987: 6) notes that four elements are included in the definition of indigenous peoples:

1. pre-existence,
2. non-dominance,
3. cultural difference, and.
4. self-identification as indigenous.

Erica-Irene Daes (2008: 17) says that four criteria can be used in the identification of indigenous peoples

1. The occupation and use of a specific territory
2. The voluntary perpetuation of cultural distinctiveness, which may include the aspects of language, social organization, religion and spiritual values, modes of production, laws and institutions
3. self-identification, as well as recognition by other groups, as a distinct collectivity

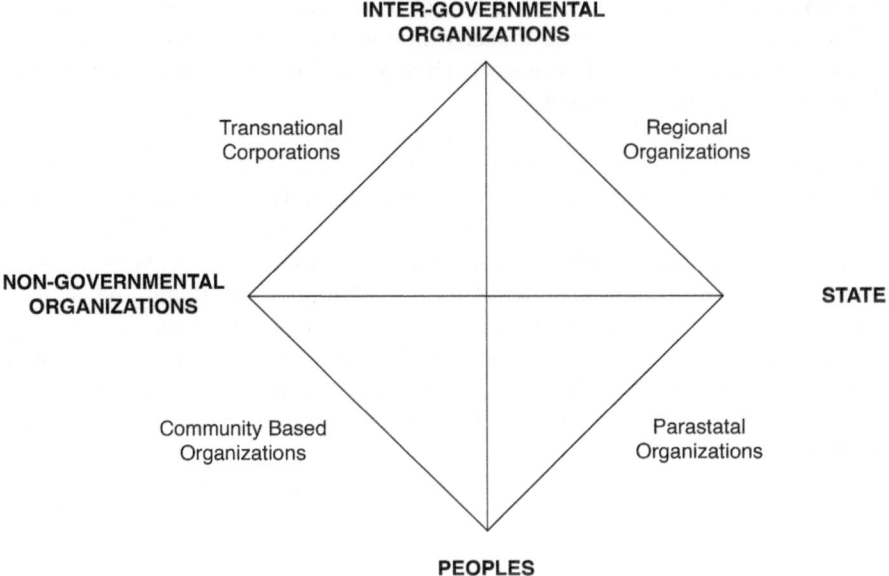

Fig. 1.1 Organization of the world system

4. an experience of subjugation, marginalization, dispossession, exclusion or discrimination

However, as Daes notes, not all four elements need to be present in order to characterize indigenous peoples. These, instead, are seen as "guiding principles."

The question remains: what are the advantages of groups identifying themselves as indigenous? From the perspective of those people who see themselves as indigenous, there are significant advantages to being identified as such. One advantage is that indigenous people can seek international assistance from the United Nations and other international organizations (IOs) and non-government organizations, including the International Work Group for Indigenous Affairs (IWGIA), Survival International, Cultural Survival, and Minority Rights Group International (Dahl, 2012; Bodley, 2015). Figure 1.1 provides and overview of the world system, which shows the various organizations that exist along with nation-state and peoples. Today, there are between 476,000,000 and 900,000,000 indigenous people residing in all of the world's continents with the exception of Antartica (Department of Social and Economic Affairs, 2009; Hall & Patrinos, 2012b; Mamo, 2022, 2023). Many of these peoples speak their own, indigenous mother-tongue languages, which together number some 5500 of a total of 700,000 of the world's languages (see Ethnologue, www.ethnologue.com, accessed 25 March 2023). There are indigenous people in over 90 of the world's 193 countries (Mamo, 2023).

Most if not all peoples who consider themselves to be indigenous or aboriginal have histories that include complex kinds of contacts with other peoples. All too often, indigenous peoples have had to cope with efforts by other groups, nation-states, settlers, or transnational corporations to take away their lands and resources, sometimes by force or through the application of questionable legal means. As David Maybury-Lewis (2002: 43) notes, 'Indigenous peoples are those who are subordinated and marginalized by those who rule over them.' Patrick Brantlinger (2003) points out that the incursions of Europeans in Australia, New Zealand, southern Africa, Latin America and North America 'meant steep population declines in indigenous populations.' He goes on to say, "One of the main causes for these declines is not mysterious: violence, warfare, genocide (Brantlinger, : 2). Another factor in the status of indigenous peoples relates to their losing access to land and natural resources and thus their livelihoods, health, and well-being.

In many instances, indigenous peoples live in places that contain a variety of kinds of valuable resources, both natural and cultural, that they wish to retain in the face of competition. There were cases where indigenous peoples actively resisted incursions by other peoples as well as assimilation and cultural modification efforts by outside agencies (Hall & Fenelon, 2009). Their cultural distinctiveness and desire to maintain their lands and identities, combined with their relative lack of power as compared to state systems, resulted in indigenous peoples being prime targets of violence or discrimination.

In most cases, indigenous peoples are ethnic minorities in multiethnic societies. The distinction drawn between indigenous peoples and minorities by the former usually revolve around (1) collective or group rights, and (2) the fact that indigenous peoples see themselves as the victims of colonization and marginalization. With very few exceptions such as Bolivia and Papua New Guinea, they do not have control over the governmental machinery of the nation-states where they live. Until relatively recently indigenous people did not have the right to vote or to take part in public policy formulation. Many members of indigenous groups believe that they lack a voice in civil affairs. They also tend to believe that they are on the bottom rung of multitiered societies. A portion of the world's indigenous societies reside in places that have come under the protection of state systems. Many of these areas are considered 'conservation hotspots' that have high ecological diversity, and which in the past were maintained by indigenous peoples who sought to ensure that the resources there were maintained and utilized in a sustainable way.

There is a long-standing debate about the ways in which indigenous peoples utilize their environments and whether their activities are sustainable (see Durning, 1992; Alcorn, 1993; Redford & Stearman, 1993a, b; Redford & Robinson, 1985; Redford & Fearn, 2007; Hames, 2007). These matters will be discussed in the balance of this book.

Types of Protected Areas

In order to understand some of the variability in types of protected areas, we have provided a list below of some of the definitions of these areas. These have been drawn from the categories provided by the International Union of the Conservation of Nature and Natural Resources.

National Park   A national park (IUCN Category II) is an area that is sizable and has a high level of protection. National parks consist of land set aside to protect outstanding natural and scenic areas of national or international significance for scientific, educational, and recreational use. These areas are relatively large natural areas not substantially altered by human activity, and where commercial extractive uses generally are not permitted. As will be shown, however, some national parks have allowed for extractive industries to operate, one example being Kakadu National Park in Australia (Table 1.1).

Strict Nature Reserve   A strict nature reserve (IUCN Category Ia) is an area which is protected from all but light human use in order to preserve the geological and geomorphological features of the region and its biodiversity. These areas are often home to ecosystems that are restricted from all human disturbance outside of scientific study, environmental monitoring. Limited recreation, and education. Because these areas are so strictly protected, they provide ideal pristine environments by which external human influence can be measured. In some cases strict nature reserves are of spiritual significance for surrounding communities, and the areas are also protected for this reason. This is the casefor example, with Bear Butte in South Dakota in the United States. The people engaged in the practice of their faith within the region have the right to continue to do so, providing it aligns with the area's conservation and management objectives.

Wilderness Area   A wilderness area (IUCN Category Ib) is similar to a strict nature reserve, but generally is larger and protected in a slightly less stringent manner. These areas are a protected domain in which biodiversity and ecosystem processes are allowed to flourish or experience restoration if previously disturbed by human

Table 1.1 Numbers of National Parks in the World

Region of world	Number of countries	Number of National Parks
Africa	54	434
Asia and the Middle East	38	895
Europe and former USSR	46	477
North America	2	110
Oceania	9	717
South America	27	403
	176 (of 193 total)	2636

Note: Data from the International Union for the Conservation of Nature and Natural Resources, the World Conservation Monitoring Center, the United Nations Evironment Program, and the World Parks Data Base activity. These are areas which may buffer against the effects of climate change and protect threatened species and ecological communities.

National Monument or Feature A natural monument or feature (IUCN Category III) is a comparatively smaller area that is specifically allocated to protect a natural monument and its surrounding habitats. These monuments can be natural in the widest sense or include elements that have been influenced or introduced by humans. The latter should hold biodiversity associations or could otherwise be classified as a historical or spiritual site, though this distinction can be quite difficult to ascertain. In order to be categorised as a natural monument or feature by IUCN's guidelines, the protected area could include natural geological or geomorphological features, culturally-influenced natural features, culturally significant sites, or cultural sites with associated landscapes. The classification then falls into two subcategories: those in which the biodiversity is uniquely related to the conditions of the natural feature and those in which the current levels of biodiversity are dependent on the presence of the sacred sites that have created an essentially modified ecosystem.

Habitat or Species Management Area A habitat or species management area (IUCN Category IV) is similar to a natural monument or feature, but focuses on more specific areas of conservation (though size is not necessarily a distinguishing feature), like an identifiable species or habitat that requires continuous protection rather than that of a natural feature. These protected areas are supposed to be controlled sufficinetly to ensure the maintenance, conservation, and restoration of particular species and habitats—possibly through traditional means. Public education in such areas is widely encouraged as part of the management objectives. Habitat or species management areas may exist as a fraction of a wider ecosystem or protected area and may require varying levels of active protection. Management measures may include (but are not limited to) the prevention of poaching, creation of artificial habitats, halting natural succession, and supplementary feeding practices of wildlife Human visitation is limited to a minimum, often allowing only those who are willing to travel of their own devices (by foot, by ski, or by boat), but this offers a unique opportunity to experience wilderness that has not been interfered with. Wilderness areas can be classified as such only if they are devoid of modern infrastructure, though they allow human activity to the level of sustaining indigenous groups and their cultural and spiritual values within their wilderness-based lifestyles.

Protected Landscape A protected landscape or protected seascape (IUCN Category V) covers an entire body of land or ocean with an explicit natural conservation plan, but usually also accommodates a range of for-profit activities. The main objective is to safeguard regions that have built up a distinct and valuable ecological, biological, cultural, or scenic character. In contrast with previous categories, Category V permits surrounding communities to interact more with the area, contributing to the area's sustainable management and engaging with its natural and cultural heritage. Landscapes and seascapes that fall into this category represent an integral balance between people and nature and can sustain activities such as traditional agricultural and forestry systems on conditions that ensure the continued protection or ecological restoration of the area. Category V is one of the more flexible classifications of protected areas. As a result, protected landscapes and seascapes may be able to accommodate contemporary developments, such as ecotourism, at the same time as maintaining the historical management practices that may procure the sustainability of agrobiodiversity and aquatic biodiversity.

Indigenous Protected Area Indigenous Protected Areas (IPAs) are found in Australia. They are managed by their traditional Aboriginal Owners in line with biodiversity consrvation agreements worked out with the Australian government (Ross et al., 2009; Gould et al., 2021). There were over 90 indigenous protected areas in Australia as of 2023 (National Indigenous Australia Agency, accessed 25 March 2023). Indigenous Protected and Conservation Areas are also found in South America and Asia.

Indigenous People in Protected Areas

People in protected areas often feel that they are at the mercy of governments, scientists, and park personnel. 'We feel as though we are pawns to power,' as a Blackfoot Indian woman said in interviews about the removals of Blackfeet and other Indians from Glacier National Park in Montana carried out in January of 2019 by Hitchcock. Similar remarks have been heard from indigenous people in the Greater Yellowstone Ecosystem (GYE), the Greater Etosha Ecosystem (GEY), the Central Kalahari Game Reserve in Botswana, and in communities surrounding Hwange National Park in Zimbabwe. The conflicts between indigenous people and governments, scientists, and park personnel are also found particularly on protected area edges (Wittemyer et al., 2008).

The social impacts of protected area establishment have been documented in detail (see Cernea & Schmidt-Soltau, 2006; West et al., 2006; Schmidt-Soltau & Brockington, 2007). There is an on-going debate, however, as to the extent to which indigenous peoples have been relocated out of protected areas and what the actual impacts have been (Redford & Fearn, 2007; Duffy, 2010; Duffy et al., 2019).

The latter years of the twentieth century and the early part of the twenty-first century have been characterized by what might be described as "eco-politics," with indigenous organizations and conservation NGOs attempting to promote environmental justice, purposely linking human rights and the environment (Durning, 1992; Rich, 1994, 2013; Sachs, 1995, 1996; Blühdorn & Welsh, 2007; Blühdorn, 2015; Johns, 2019; Mamo, 2022). Environmental ethics and social justice are being pushed by both social and environmental scientists and development workers (see, for example, Kopnina, 2013; Kopnina & Shoreman-Ouinet, 2015; Shoreman-Ouinet & Kopnina, 2015; Kopnina & Washington, 2020). Indigenous and environmental groups have attempted to enhance their impacts through their collaborative efforts, which ultimately could lead to the international recognition of a communal right to a healthy environment. As part of these discussions, it has been claimed that good-quality habitats with diverse species are crucial to human and other species' well-being. It has been shown that areas that are in the hands of indigenous people tend to be managed better and have less evidence of deforestation and environmental degradation than other areas (see, for example, Alves-Pinto et al., 2022).

A primary reason for the increased awareness of the links among human rights, social justice, and the environment is the rapid expansion in the exploitation of natural resources and extablishment of extractive industries in some of the planet's remotest, and potentially richest, habitats as a result of globalization. Globalization is a process through which communities, organizations, nation-states become increasingly interconnected economically, socially, and technologically due to improvements and changes in communication systems and the expanded international trade of goods and services. One outgrowth of globalization has the rapid movements of people, goods, and information across the globe (Wallerstein, 1979). Another outgrowth of globalization is an increase in the gap between individuals, communities, and nation-states that are well off and those that are poor (Goldman, 2006; Rudra, 2008; Hall & Fenelon, 2009; Walter, 2021). The promises of globaliztion to increase overall wealth among the poor have not come to fruition.

As noted in this volume, one set of groups on the planet that have been impacted especially negatively by globalization and modernization processes is indigenous peoples, in part because many of them reside in areas that have high concentrations of valuable resources, including minerals, oil, timber, and non-timber forestry products (Durning, 1992; Department of Social and Economic Affairs, 2009; Hall & Patrinos, 2012a, b). Some of the most vulnerable indigenous peoples are those living in voluntary isolation in places ranging from the Amazon to central Africa and from the forests of southeast Asia to the islands making up the Andamans in the Bay of Bengal and the islands of Indonesia, Malaysia, and the Philippines. Organizations and policies have been established to protect people in voluntary isolation and stages of initial contact, as seen, for example, in the Amazon Basin (Zarzar, 2000; Huertas Castillo, 2004, 2008; Defensorı'a del Pueblo, 2006; Napolitano, 2007).

Indigenous peoples, as will be shown, have extensive indigenous and traditional knowledge which they draw upon in adapting to their environments (Popova-Gosart, 2009; Magni, 2017; Tom et al., 2019; Salim et al., 2023). In many cases, indigenous groups share that knowledge with each other, and they speak of the importance of indigenous or traditional knowledge in addressing complex issues such as climate change, as seen, for example, among the San of southern Africa (see, for example, Backwell & D'Errico, 2021). Indigenous knowledge of plants, insects, and other species is significant and is maintained by indigenous populations such as the G/ui and G//ana San of Botswana (see Sapignoli, 2018) and the Ju/'hoansi of Namibia and Botswana (Hitchcock, 2022, 2023; Leon Tsamkxao, personal communication, 24 March 2023).

Indigenous people have sought to have a say in international conferences dealing with the environment and climate change, the most recent one being COP 27, held in Sharm-el Shaik in Egypt in November, 2022. While they were able to attend some of the sessions at the meeting, they had little, if any, input in the final reports and recommendations of the meeting (Bixcul, 2023). In the past, indigenous peoples were left out of many of the international environmental meetings, going back to the UN Conference on Environment and Development (UNCED) held in Rio de Janeiro in 1992 (Spector et al., 1994). A trend in the past 30 years has been an expansion in the number of indigenous people and indigenous organizations attending international environmental and human rights conferences.

In this book, we focus specifically on indigenous peoples around the world, particularly those who live in areas that have become protected as national parks, game reserves, forestry reserves, national monuments, and wilderness areas. We begin with a discussion of what has come to be called fortress conservation or militarized conservation, examining a set of cases in the United Sates (Chap. 2). We then go on to discuss coercive conservation as it has occurred in southern Africa (Chap. 3). After that, we assess some of the impacts of conservation-related involuntary resettlement (Chap. 4). Next, we assess strategies employed by indigenous peoples aimed at ensuring their legal rights to land and resources (Chap. 5). We conclude with a discussion of the ethics of conservation and resettlement with particular reference to indigenous peoples (Chap. 6). We hope in this volume that the complex relationships among people, parks, and power are demonstrated.

References

Alcorn, J. (1993). Indigenous peoples and conservation. *Conservation Biology, 7*(2), 424–425.

Alves-Pinto, H. N., Cordeiro, C. L. O., Geldmann, J., Jonas, H. D., Gaiarsa, M. P., Balmford, A., Watson, J. E. M., Latawiec, A. E., & Strassbur, B. (2022). The role of different governance regimes in reducing native vegetation conversion and promoting regrowth in the Brazilian Amazon. *Biological Conservation, 267*, 109473. https://doi.org/10.1016/j.biocon.2022.109473

Anaya, S. J. (2009). *Indigenous peoples in international law*. Wolters Kluwer.

Backwell, L., & D'Errico, F. (2021). *San elders speak: Ancestral knowledge of the Kalahari San*. Wits University Press.

Bixcul, B. (2023). Despite historic participation indigenous peoples are again sidelined in major decisions at COP 27. *Cultural Survival Quarterly, 47*(1), 12–15.

Blühdorn, I., & Welsh, I. (2007). Eco-politics beyond the paradigm of sustainability: A conceptual framework and research agenda. *Environmental Politics, 16*(2), 185–205.

Blühdorn, I. (2015). *A much-needed renewal of environmentalism? Eco-politics in the Anthropocene*. Routledge.

Borgerhoff Mulder, M., & Coppolillo, P. (2005). *Conservation: Linking ecology, economics, and culture*. Princeton University Press.

Bodley, J. H. (2015). *Victims of progress* (6th ed.). AltaMira Press.

Brantlinger, P. (2003). *Dark vanishings: Discourse on the extinction of primitive races 1800–1930*. Cornell University Press.

Brysk, A. (2000). *From tribal village to global village: Indian rights and international relations in Latin America*. Stanford University Press.

Cernea, M., & Schmidt-Soltau, K. (2006). Poverty risks and National Parks: Policy issues in conservation and resettlement. *World Development, 34*(10), 1808–1830.

Daes, E.-I. (2008). *Indigenous peoples: Keepers of our past, custodians of our future*. International Work Group for Indigenous Affairs.

Dahl, J. (2012). *The indigenous space and marginalized peoples in the United Nations*. Palgrave Macmillan.

Defensoría del Pueblo. (2006). *Informe No. 101: Pueblos indígenas en situación de aislamiento voluntario y contacto inicial* [Report No. 101: *Indigenous people in voluntary isolation or initial contact*]. Defensoría del Pueblo.

Department of Social and Economic Affairs. (2009). *State of the world's indigenous peoples*. Department of Social and Economic Affairs, United Nations.

Dowie, M. (2005). Conservation refugees: When protecting nature means kicking people out. *Orion, 24*(6), 16–27.

Dowie, M. (2009). *Conservation refugees: The hundred year conflict between global conservation and native peoples*. MIT Press.

Duffy, R. (2010). *Nature crime: How we're getting conservation wrong*. Yale University Press.

Duffy, R., Massé, F., Smidt, E., Marijnen, E., Büscher, B., Verweijen, J., Ramutsindela, M., Simlai, T., Joanny, L., & Lunstrum, E. (2019). Why we must question the militarization of conservation. *Biological Conservation, 232*, 66–73.

Durning, A. T. (1992). *Guardians of the land: Indigenous peoples and the health of the Earth*. WorldWatch Paper 112. WorldWatch Institute.

Fanari, E. (2019). Relocation from protected areas as a violent process in the recent history of biodiversity conservation in India. *Ecology, Economy, and Society – the ENSSE Journal, 2*(1), 43–76.

Goldman, M. (2006). *Imperial nature: The World Bank and struggles for social justice in the age of globalization*. Yale University Press.

Gould, J., Smyth, D., Rassip, W., Rist, P., & Oxenham, K. (2021). Recognizing the contribution of indigenous protected areas to marine protected area management in Australia. *Maritime Studies, 20*(1), 5–26.

Hall, G., & Patrinos, H. A. (2012a). Latin America. In G. Hall & H. A. Patrinos (Eds.), *Indigenous peoples, poverty, and development* (pp. 344–358). Cambridge University Press.

Hall, G., & Patrinos, H. A. (Eds.). (2012b). *Indigenous peoples, poverty, and development*. Cambridge University Press.

Hall, T. D., & Fenelon, J. V. (2009). *Indigenous peoples and globalization: Resistance and revitalization*. Paradigm Publishers.

Hames, R. B. (2007). The ecologically noble savage debate. *Annual Reviews of Anthropology, 36*, 177–190.

Hitchcock, R. K. (2022). Kalahari San valuation of nature. In *Intergovernmental Platform on Biodiversity and Ecosystem Services (IPBES) Values Assessment*. Intergovernmental Platform on Biodiversity and Ecosystem Services (IPBES), Institute for Ecosystems and Sustainability Research.

Hitchcock, R. K. (2023). Climate change resilient livelihoods and adaptive strategies among the Ju/'hoansi San of Nyae Nyae, Namibia. In P. Gadhoke, B. P. Brenton, & S. Katz (Eds.), *Global transformations of food systems for climate change resilience* (pp. 102–134). CRC Press.

Huertas Castillo, B. (2004). *Indigenous peoples in isolation in the Peruvian Amazon*. International Work Group for Indigenous Affairs (IWGIA).

Huertas Castilo, B. (2008). *Los Pueblos Indigenas en Aislamiento: Su Luchu por la Sobrevivencia y la Libertrad* [Indigenous peoples in isolation: Their struggle for survival and liberty]. International Work Group for Indigenous Affairs.

Igoe, J., & Brockington, D. (2006). Eviction for conservation: A global overview. *Conservation and Society, 2*(2), 411–432.

Independent Commission on International Humanitarian Issues. (1987). *Indigenous peoples: A global quest for justice*. Zed Books.

Johns, D. (2019). *Conservation politics: The last anti-colonial Battle*. Cambridge University Press.

Kopnina, H. (2013). Forsaking nature? Contesting 'biodiversity' through competing discourses of sustainability. *Journal of Education in Sustainable Development, 9*(4), 235–254.

Kopnina, H. (2016). Half the earth for people (or more)? Addressing ethical questions in conservation. *Biological Conservation, 203*, 176–185.

Kopnina, H., & Shoreman-Ouinet, E. (2015). *Sustainability: Key issues*. Routledge.

Kopnina, H., & Washington, H. (Eds.). (2020). *Conservation: Integrating social and ecological justice*. Springer.

Kuper, A. J. (2003). The return of the native. *Current Anthropology, 44*(3), 389–411.

Magni, G. (2017). Indigenous knowledge and implications for the sustainable development agenda. *European Journal of Education, 52*, 437–447.

Mamo, D. (Ed.). (2022). *The indigenous world 2022*. International Work Group for Indigenous Affairs.

Mamo, D. (Ed.). (2023). *The indigenous world 2023*. International Work Group for Indigenous Affairs.

Maybury-Lewis, D. (2002). Genocide against indigenous peoples. In A. L. Hinton (Ed.), *Annihilating difference: The anthropology of genocide* (pp. 43–53). University of California Press.

Merlan, F. (2009). Indigeneity: Local and global. *Current Anthropology, 50*(3), 202–333.

Mushonga, T., & Matose, F. (2020). Dimensions and corollaries of violence in Zimbabwe's protected forests. *Geoforum, 117*, 216–224.

Napolitano, D. A. (2007). Towards understanding the health vulnerability of indigenous peoples living in voluntary isolation in the Amazon Rainforest: Experiences from the Kugapakori Nahua Reserve, Peru. *EcoHealth, 4*, 515–531.

Naughton-Treves, L., Holland, M. B., & Brandon, K. (2005). The role of protected areas in conserving biodiversity and sustaining local livelihoods. *Annual Review of Environment and Resources, 30*, 219–252.

Niezen, R. (2003). *The origins of Indigenism: Human rights and the politics of identity*. University of California Press.

Owens, M., & Owens, D. (1981). *Preliminary final report on the central Kalahari predator research project*. Report to the Department of Wildlife and National Parks, Gaborone, Botswana.

Owens, M., & Owens, D. (1984). *Cry of the Kalahari*. Houghton-Mifflin.

Popova-Gosart, U. (2009). *Traditional ecological knowledge and indigenous peoples*. Gorno-Altaisk, Altai Republic, Russian Federation: L'auravetl'an Information and Education Network of Indigenous Peoples (LIENIP) and Geneva: World Intellectual Property Organization (WIPO).

Ramutsindela, M., Matose, F., & Mushona, T. (Eds.). (2022). *The violence of conservation in Africa: State, militarization, and alternatives*. Edward Elgar Publishing.

Redford, K., & Robinson, J. (1985). Hunting by indigenous peoples and conservation of game species. *Cultural Survival Quarterly, 9*(1), 41–44.

Redford, K. H., & Stearman, A. M. (1993a). Forest dwelling-native Amazonians and the conservation of biodiversity: Interests in common or collusion? *Conservation Biology, 7*(2), 248–255.

Redford, K. H., & Stearman, A. M. (1993b). On common ground? Response to Alcorn. *Conservation Biology, 7*(2), 248–255.

Redford, K., & Fearn, E. (Eds.). (2007). *Protected areas and human displacement: A conservation perspective*. Working paper no. 29. Wildlife Conservation Society.

Rich, B. (1994). *Mortgaging the earth: The World Bank, environmental impoverishment, and the crisis of development*. Beacon Press.

Rich, B. (2013). *Foreclosing the future: The World Bank and the politics of environmental destruction*. Island Press.

Ross, H., Grant, C., Robinson, C. J., Izurieta, A., Smyth, D., & Rist, P. (2009). Co-management and indigenous protected areas in Australia: Achievements and ways forward. *Australasian Journal of Environmental Management, 16*(4), 242–252.

Rudra, N. (2008). *Globalization and the race to the bottom in developing countries: Who really gets hurt?* Cambridge University Press.

Sachs, A. (1995) *Eco-justice: Linking human rights and the environment*. WorldWatch Paper 127. WorldWatch Institute.

Sachs, A. (1996). Upholding human rights and environmental justice. In L. Brown et al. (Eds.), *State of the World 1996* (pp. 133–151). WorldWatch Institute.

Salim, M. J., Anuar, S. N., Omar, K., Mohamad, T. R. T., & Sanusi, N. A. (2023). The impacts of traditional ecological knowledge towards indigenous peoples: A systematic literature review. *Sustainability, 15*, 824. https://doi.org/10.3390/su15010824

Sapignoli, M. (2018). *Hunting justice: Displacement, law, and activism in the Kalahari. Law and society series*. Cambridge University Press.

Schmidt-Soltau, K., & Brockington, D. (2007). Protected areas and resettlement: What scope for voluntary relocation? *World Development, 35*(12), 2182–2202.

Shoreman-Ouinet, E., & Kopnina, H. (2015). Reconciling ecological and social justice to promote biodiversity. *Biological Conservation, 184*, 320–226.

Spector, Bertram 1. Spector, Gunnar Sjostedt, and 1. William Zartman. (1994). *Negotiating international regimes: Lessons learned from the United Nations conference on environment and development (UNCED)*. Martinus Nijhof.

Tauli-Corpuz, V., Alcorn, J., & Molnar, A. (2018). *Cornered by protected areas: Replacing 'fortress' conservation with rights-based approaches helps bring justice for indigenous peoples and local communities, reduces conflict, and enables cost-effective conservation and climate action*. Rights and Resources Initiative.

Tauli-Corpuz, V., Alcorn, J., Molnar, A., Healy, C., & Barrow, E. (2020). Cornered by PAs: Adopting rights-based approaches to enable cost-effective conservation and Climate Action. *World Development, 130*. https://doi.org/10.1016/j.worlddev.2020.104923

Tom, M. N., Huaman, E. S., & McCarty, T. L. (2019). Indigenous knowledges as vital contributions to sustainability. *International Review of Education, 65*, 1–18.

Wallerstein, I. (1979). *The capitalist world economy*. Cambridge University Press.

Walter, S. (2021). The backlash against globalization. *Annual Review of Political Science, 24*, 421–442.

West, P., Igoe, J., & Brockington, D. (2006). Parks and peoples: The social impact of protected areas. *Annual Review of Anthropology, 35*, 251–277.

Wilson, E. O. (1988). The current state of biological diversity. In E. O. Wilson (Ed.), *Biodiversity* (pp. 3–18). National Academy Press.

Wilson, E. O. (2016). *Half-earth: Our planet's fight for life*. Livewright Publishing Corporation.

Wittemyer, G., Elsen, P., Bean, W. T., Burton, C. C., & Brashares, J. S. (2008). Accelerated human population growth at protected area edges. *Science, 321*, 123–126.

Workman, J. G. (2009). *Heart of dryness: How the last bushmen can help us endure the coming age of permanent drought*. Walker and Company.

Zarzar, A. (2000). *Tras las huellas de un antiguo presente: la problema'tica de los pueblos amazo'nicos en aislamiento y contacto inicial* [In the tracks of an ancient present: The situation of Amazonian peoples in isolation and initial contact]. Defensori'a del Pueblo.

Zips-Mairitsch, M. (2013). *Lost Land? (Land) Rights of the San in Botswana and the legal concept of indigeneity in Africa*. Lit Verlag and Copenhagen: International Work Group for Indigenous Affairs.

Chapter 2
Fortress Conservation: Removals of Indigenous People from Protected Areas in the United States

Introduction

This chapter considers examples of removals of people from protected areas in the United States. We start out with the best-known national park in the world—and the first one established on the planet—Yellowstone National Park in the United States. We then examine two other early national parks in the United States, Yosemite National Park and Glacier National Park. Table 2.1 presents data on three national parks and a national historic park in the United States. After that we consider a series of examples in the Southwestern United Sates where indigenous people were required to relocate from their ancestral lands. Nearly all of these protected areas have witnessed coercive conservation strategies, some of which involved militaries or state agencies that employed complex means to remove people from areas that were set aside as protected. Each of the cases contains a discussion of the indigenous people who were affected. The issue of tourism is also considered.

Case 1: Yellowstone National Park in the United States

In order to assess the consequences of resettlement out of protected areas, we look at several examples. The first of these is Yellowstone National Park in the United States. Yellowstone National Park was established in the United States on 1 March, 1872, the world's first national park. This park, which is part of three American states (Wyoming, Montana, and Idaho) was associated with 26 Native American tribes. Yellowstone, unlike another national park founded in California some years later, Yosemite, which was established on 1 October, 1890, saw removals of all of the American Indian groups that either resided in or utilized the resources in the park. Some of this was done by agents of the US government, notably the U.S. military and later park rangers, beginning in 1877. The reasons for the disposession were,

© The Author(s) 2023
M. Sapignoli, R. K. Hitchcock, *Anthropology and Ethics*,
https://doi.org/10.1007/978-3-031-39268-9_2

Table 2.1 National Parks in the United States and their affiliated tribes

Name and state	Date of establishment	Size (km²)	Number of affiliated tribes
Glacier National Park, Montana	1910 (10 May)	4100.77 km²	14
Kaloko-Honokohau National Historic Park, Hawaii	1978 (November 10)	1030 acres; 4.17 km²	2
Yellowstone National Park	1872 (1 March)	8983.18 km²	26
Yosemite National Park	1890 (October 1)	3027 km²	7
Totals	4	16,025.12 km²	49

Note: Data obtained from the National Park Service, U.S. Department of the Interior, Washington D.C. and from the websites of the individual monuments and national parks and specific tribes

according to staements made by government personnel at the time, (1) to protect the 'wilderness' aspects of the park and (2) to ensure the safety of tourists (Spence, 1996a: 38–39; Loendorf & Stone, 2006).

The list of the 26 tribes associated with Yellowstone National Park are presented below: Assiniboine and Sioux Tribes, Blackfeet Tribe, Cheyenne River Sioux Tribe, Comanche Tribe of Oklahoma, Confederated Salish and Kootenai Tribe, Confederated Tribes of the Colville Indian Reservation, Confederated Tribes of the Umatilla Indian Reservation, Couer d'Alene Tribe, Crow Tribe, Crow Creek Sioux Tribe, Eastern Shoshone Tribe, Flandreau Santee Sioux Tribe Gros Ventre and Assiniboine Tribe, Kiowa Tribe of Oklahoma, Lower Brule Sioux Tribe, Nez Perce Tribe, Northern Arapaho Tribe, Northern Cheyenne Tribe, Oglala Sioux Tribe, Rosebud Sioux Tribe, Shoshone-Bannock Tribe, Siseton-Wahpeton Sioux Tribe, Spirit Lake Sioux Tribe, Standing Rock Sioux Tribe, Turtle Mountain Band of the Chippewa Indians, and Yankton Sioux Tribe.

American Indian tribes in Yellowstone that were affected by the US government policies included the so-called Sheepeaters, a subgroup of Shoshones, the Nez Perce, who traveled from Idaho through Yellowstone in 1877, and the Bannocks, who entered Yellowstone to obtain horses in 1878. In 1882, there were negotiations that led to the Sheepeaters leaving the park and moving to reservations in Idaho and Montana (Spence, 1996a, b: 39; Loendorf & Stone, 2006). In 1886, the U.S. military was granted control of Yellowstone, who concentrated their efforts on removing the Bannocks, Sheepeaters, and Shoshones (Spence, 2000: 62–66). Indians still entered the park periodically to hunt and to obtain wild natural resources, including lodgepole pines for ceremonial purposes. They also employed fire to manage the resources in Yellowstone, something done by native peoples in many areas of North America and a subject of extensive discussion about indigenous resource management. The U.S.military focused its attention on Indian hunting activities and fire-setting after their arrival in 1886, which lasted into the early twentieth century.

Yellowstone National Park, which today is 8983.18 km² in size, can be viewed in several ways (Black, 2012; Farrell, 2015). It is the oldest national park in the world, having been established in 1872. Considered as 'America's wild idea' (Quammen, 2018) or "America's best idea' (Ken Burns, who did a 6-part film on US national parks in 2009). Yellowstone is in many ways a paradox. It provides 'benefits for the nation' while at the same time it was established on the land of a sizable number of indigenous people who were displaced without compensation (Spence, 1996a, b, 2000; Janetski, 1987; Nabokov & Loendorf, 2004; Loendorf & Stone, 2006). If one goes to the park, there is little evidence in the main visitor center that suggests that the indigenous residents were forcefully relocated. Yellowstone National Park does, however, have some books in its bookstore that deal with Indian issues. On the other hand, there is substantial coverage of wildlife issues in and around the park—one of the last free-ranging herds of bison is found in Yellowstone and its surroundings (Meagher, 1973; Beschta et al., 2020). The Yellowstone elk herd, which is large, migrates through the park and into buffer zones, where a protion of the elk can be shot (Dan MacNulty, personal communication, October 2021). The management of the Yellowstone elk herd is both complex and innovative (Grindle, 2023).

According to the National Park Service, there are over 1800 archaeological and 1300 sites of cultural and natural significance in Yellowstone, a park that today covers 8983.18 km^2 (3472 mi^2). In the past, various Indian peoples visited the hot springs and other volcanic features in Yellowstone. Indigenous people obtained obsidian (volcanic glass) from Obsidian Cliffs for making arrowheads and other technological items (Cannon, 1993; MacDonald, 2018). Some tribes saw the highest point in Yellowstone, Mount Eagle (1610 m in elevation) as a sacred place, and they fished in Lake Yellowstone and the rivers and creeks in the area. They also hunted a wide variety of animals in the Greater Yellowstone Ecosystem (Nabokov & Loendorf, 2004:108–117, 169–173, 208–210; Loendorf & Stone, 2006: 58–59, 140–145; MacDonald, 2018: 36–37, 50–51, 60, 73, 117–120, 169–171, 172–178).

Indigenous peoples foraging in this 'wild' area occupied the ecological niche of an apex/keystone predator whose presence would have significantly impacted the population density of local prey species (Kay, 2007). Moose were the most susceptible to aboriginal hunting followed by bison and then elk (Kay, 2007: 1). Indigenous people employed a variety of strategies to manage the natural resources in Yellowstone: fire was uses to burn off areas in order to maintain ecological integrity, and selective exploitation of plant species took place. Hunting was done carefully, with efforts to diversity the exploitation of wild animals both in a spatial and temporal sense. See also Chacon and Mendoza (2012) who likewise stress the importance of recognizing the impact that native peoples had (and continue to have) in shaping local ecosystems.

In general, far more attention appears to be paid in Yellowstone National Park to the animal biodiversity in the protected area, particularly the controversial issue of gray wolves (*Canis lupus*) and their re-introduction (Bangs et al., 2008; Smith & Ferguson, 2012; Ruth et al., 2019). Wolves were re-introduced in Yellowstone in June of 1995, the original wolves coming from Canada (Duffield et al., 2006). In late 2018, there were 96 wolves in 10 packs in Yellowstone. Wolf tourism in the park brought $35 million in 2018 (Dan MacNulty, personal communication, 2019). As Doug Smith noted in a *60 Minutes* segment broadcast in December, 2018, 'Wolves are the saviors of Yellowstone ecology.' He pointed out that the power of wolves is in the pack and the way it operates (for further discussions of the impacts of wolf reintroduction, see Smith & Ferguson, 2012; Smith & Peterson, 2021). Wolves were a keystone species in Yellowstone in the 1700s, and they had significant effects on the prey in the Greater Yellowstone Area (GYA) until they were eliminated in the twentieth century. Wolf translocation has led to an increase in the number of wolves in Yellowstone, which had increased to 120 in 2023.

Surrounding Yellowstone Park in the states of Montana, Idaho, and Wyoming, there are people who both love and hate wolves. Ranchers who have cattle and sheep, for example, tend to dislike wolves, and have sought rights to kill them. Today, there are some 3123 wolves in the three states. Wolf hunting has been reintroduced in Montana. As Smith pointed out, 'People feel better if they are not powerless' and if they have some say over decision-making about wild animals like wolves." Wolf relationships with elk and other species are important, and they affect the population size, population dynamics, and movement patterns of elk. It is interesting to note that hunters on the peripheries of the park play a significant role in terms of affecting the numbers of elk and moose in the Greater Yellowstone Area. Wolf recovery has been an important area of discussion in Yellowstone (Smith et al., 2020a, b), and it continues to be controversial. Also controversial is the fact that gold mining is done in the Yellowstone River at Gardiner, Montana, by a Canadian gold mining company (Soave, 2018). The issue of extractive indusries in and around protected areas is a cricial one.

The animal that has received the greatest scientific attention in the Greater Yellowstone Ecosystem is the buffalo (*Bison bison*). (Callenbach, 1995; Geist, 1997). Bison have a long and compex history in the GYE (Cannon et al., 2023). Mobility patterns have changed, with some bison remaining in basins and others migrating into the uplands and back down into lower elevations. Bison hunting by Native Americans, inside and outside of Yellowstone up until the removals occurred has varied over time (Meagher, 1973; Isenberg, 2000). One of the concerns of ranchers living on the peripheries of Yellowstone is that the bison carry a disease, brucellosis, which affects cattle.

In the US, especially after June 25, 1876 and the Battle of the Little Bighorn in Montana and the defeat of General George Armstrong Custer and the Seventh Cavalry by Lakota, Cheyenne, and Arapaho, there were numerous statements made in Congress that Indians were dangerous and either should be eliminated completely or moved to reservations where they could be under the watchful eye of government authorities (Utley, 1984; Anderson, 2014). Later on, particularly in the late 1870s, 1880s and 1890s, it was argued by US government officials that Indians were harmful to wilderness areas (Spence, 1996a: 40). The Yellowstone model of 'Fortress Conservation" as it has been described (see Brockington, 2002), was predicated on the notion that conservation could only be successful if the indigenous inhabitants and users of these areas were removed.

One of the trends in the twenty-first century has been the effort to 're-indigenize' national parks and protected areas in the United States (Treuer, 2021; Stark et al., 2022). What this means is that indgenous peoples should have the right to have a say about the ways in which national parks and protected areas are managed. Members of the American Indian Movement (AIM) feel that native people should have the right to re-occupy national parks and monuments, one example being the take-over of Alcatraz Island by Indian activists in November 1969 and another the take-over of Wounded Knee, South Dakota in 1973 (Smith & Warrior, 1997; Treuer, 2019). A goal of Native American activists has been the return of protected areas to indigenous people so that they can manage them the way that they wish, that is, using sustainable methods.

Case 2: Yosemite National Park, United States

In the western United States were relocated to Indian reservations in the 1870s through the end of the nineteenth century. It is important to point out that Yosemite in the western Sierra Nevada Mountains of California was originally conceived as an 'Indian-free park' on the recommendation of conservationist and Sierra Club Founder John Muir in 1890. The residents of what was to become Yosemite were the Ahwahneechee, which, when translated, means "dwellers in Ahwahnee' or, in Southern Miwok, 'gaping mouth-like place' (Greene, 1987; Runte, 1990). The Ahwahneechee lived and obtained resources in what was to become Yosemite National Park for dozens of generations. Initial contacts between indigenous people in Yosemite and outsiders were fleeting, but as non-Indians flooded in to California after the discovery of gold at Sutter's Mill in 1849, the contacts became much more frequent. Some of these contacts were peaceful, consisting of exchanges of goods, but others were more violent, resulting in injuries and deaths of both indigenous and non-indigenous people (Fig. 2.1).

After a number of violent encounters between whites and Indians in northern California, a militia was formed, known as the Mariposa Battalion, in 1851 which went into the Yosemite Valley and attacked the residents under Chief Taneya (Madley, 2016: 186–194). The 200 or so Ahwahneechee who survived were removed to the Fresno River reservation (Leshy, 2021: 101), but some of them were allowed to return to areas close to Yosemite. Further conflicts ensued, resulting in additional efforts at removal which ultimately were usuccessful. There were periodic vigilante and militia campaigns into the Yosemite area which led to the revenge killings of Indians from various groups (Madley, 2016: 193–194). Several groups of Indians remained in and around Yosemite, working in various capacities, including for the National Park Service after its formation in 1916, and they interacted with park personnel and visitors, selling crafts such as baskets and demonstrating traditional lifeways until 1969, when the village of Wahhoga at Yosemite National Park was finally closed down. Indian Field Days were held in Yposemite beginning in August of 1916 and versions of these continue to the present.

There are seven affiliated tribes who are associated with Yosemite National Park and who have ancestral connections to the park. These include the following: Bishop Paiute Tribe, Bridgeport Indian Colony (Yosemite Paiutes), Mono Lake Kootzad'ka'a Tribe, North Fork Rancheria of Mono Indians of California, Picayune

Fig. 2.1 Western United States National Parks. Source: Abercrombie and Kent

Rancheria of the Chukchansi Indians, Southern Sierra Miwuk Nation, and Tuolomne Band of Mi-wuk Indians For further information on these affiliated tribes, see National Park Service (2019).

Abraham Lincoln signed an act of Congress that created a Yosmite state park in 1864 The residents of the Yosemite area were some of the first Native Americans to experience the costs and benefits of park tourism (Burnham, 2000: 20). Indians were tolerated in the early years of the state park, where they helped guide tourists and fishing parties, harvest wood and hay, and sell crafts to visitors. The incursions of miners and settlers saw repeated violent conflicts and evictions of Miwok and Ahwahneechee during what was termed the Mariposa Indian War (Madley, 2016: 189–194) in which hundreds of Indians were killed. When the Miwok and Ahwahneechee returned to the Yosemite area, they continued to manage the timber and other resources through careful fire use. The Yosemite Valley had a continuous Indian presence until well into the twentieth century (Spence, 1999: 101). They found that they could both earn a livelihood in the park and at the same time remain in the ancestral area. The remoteness of Yosemite meant that their labor was highly prized, and there were symbiotic relationships between native and non-native residents of the area. The Ahwahneechee considered the Yosemite Valley a sacred place and the non-native residents of the valley recognized its value, as well.

Yosemite, a national park which in 1984 was declared as a World Heritage Site, is 3027 km^2 in size and falls in three California counties. Historically, Yosemite, in contrast to Yellowstone and Glacier, was largely tolerant of the presence of indigenous people, at least for some parts of its history. Yosemite Park rangers were known to draw heavily on the imagery of Native American traditional life for tourism purposes. Annual consultation meetings have been held between Yosemite National Park represenatives and the consortium of tribes affiliated with the park for a decade, allowing for decisions to be made that include park administrators, representatives of the Yosemite Conservancy, and the seven affiliated tribes.

The seven affiliated tribes associated with Yosemite National Park have had some benefits from Yosemite, including employment, opportunities to work with scientists and park managers, and generation of income through participation in tourism in Yosemite. They also now have the right to visit some of the places that they considered sacred that are inside of the park. In addition, they can share their knowledge about complex matters such as climate change and how they adapted to it in the past and how they hope to deal with climate change and human-induced change in the future.

Case 3: Glacier National Park, Montana

Glacier National Park in the state of Montana in the United States was established on 11 May 1910. Glacier is 4100.77 km^2 in size. Its creation led to the eviction of indigenous peoples, notably the Blackfeet in the east and the Flathead in the west (Spence, 1996b, 2000: 71–100). Attorneys for the United States argued that the Glacier National Park Act of 1910 extinguished the Blackfeet rights to the western part of Glacier. In 1935, the Blackfoot Nation brought suit against the United States, asserting that the creation of Glacier National Park amounted to a governmental taking, as it deprived them of the right to hunt and fish on the tract of land that they had sold to the United States under the Act of June 1, 1896 (Sholar, 2004: 151). The Court of Claims ruled against the Blackfeet and abrogated their claims to lands on which to hunt and fish (Sholar, 2004: 159). In addition, the Supreme Court reasoned that before the creation of the park in 1910, the Blackfeet, "did not exercise to any appreciable extent the rights reserved in [the 1896 agreement], to hunt and fish in and remove timber from the land ceded in the agreement, and such rights were authoritatively terminated," by the creation of Glacier National Park (Sholar, 2004: 159). The Court also said that the Blackfeet were not entitled to any compensation (Blackfeet, Blood, Piegan & Gros Ventre Nations or Tribes of Indians v. United States, 81 Ct. Cl. 101, 115 (1935)). Eventually, the Blackfeet lost all rights to Glacier

National Park. In recent years, however, the Blackfeet have been negotiating with National Park Service authorities in order to obtain co-management rights over Glacier National Park, in addition to setting up a tribal park on Glacier's western boundary in Montana (Craig et al., 2012).

The relocation of the Blackfeet, Blood, Pieghan, and Flathead Indians tribes from Glacier resulted in a wide range of impacts. The hunting and gathering areas of the Indian communities were reduced in size and richness, which in turn affected their nutritional statuses. Their loss of access to sacred sites in Glacier National Park had negative effects on their belief systems and contributed, as numerous indigenous consultants said in interviews, to the decline of their psychological well-being. The relocations of various Indian groups led to social disruptions and changes in self-identity. There were also increased social disruptions and tensions that emerged among the various Indian communities who had lived in and utilized the resources of Glacier National Park. Key areas that supported Indians in the winter were the river valleys on the eastern side of the mountains.

Drawing on data from observers, some of them summarized in Spence (1995, 2000: 72–73), and the National Park Service, there were 14 groups that had associations with Glacier National Park. These include the following: Assiniboine, Blackfeet, Bloods (Kaina), Cree, Crow, Gros Ventre (Atsina), Kalispel (Pond'Orielle), Natoka (Stoney), Nitsitapi, North Piegen, South Piegen, Pikuni, Salish Koutenai (Flathead), and Siksika It is important to note that few of the Indian tribes opposed the creation of Glacier National Park while some were vocally in support of it. As time went on, however, Blackfeet were drawn upon to perform for tourists at the main park headquarters in the 1930s, much to the resentment of other tribes. It was Blackfeet who were most involved in the litigation involving the Department of the Interior and Glacial National Park authorities in the 1930s (Spence, 2000: 89–100). An on-going set of tensions, lasting until recent times, has been over game protection from Indian hunters and whites who enter the park illegally (Spence, 2000: 88–89). There were also pressures brought to bear on Glacier National Park to allow the exploitation of key resources for Indian ceremonies, including medicinal plants and dyes for sacred objects, something that the Park Authories were reluctant to allow.

Glacier National Park is now part of a transboundary protected area. Bordering on Glacier is Waterton Lakes National Park which is located in the southwestern corner of the province of Alberta, Canada. Waterton was the fourth Canadian national park when it was established in 1895. It was named after Waterton Lake, which in turn was named in honor of the Victorian naturalist and conservationist Charles Waterton. This national park contains 505 km^2 (195 mi^2) of rugged mountains, valleys, prairie, lakes, rivers, and wilderness. Operated by Parks Canada, Waterton is open throughout the year, but the main tourist season is during July and August. In 1932, Waterton Lakes National Park in Canada was combined with Glacier National Park in the United States to form the world's first International Peace Park. The size of Waterton-Glacier International Peace Park is 4576.14 km^2. The total area is sometimes referred to as the "Crown of the Continent Ecosystem," which extends over 41,000 km^2 (Spence, 1996b). Travel between Glacier National Park and

Waterton Lakes National Park was reduced in 2020–2021 as a result of restrictions imposed as a result of the Coronavirus pandemic.

Kaloko-Honokōhau National Historic Park

Kaloko-Honokōhau is a relatively small (1030 acres) national historic park lying midway between Kailua-Kona and the Keahole International Airport on the leeward side of the Big Island of Hawaii. Kaloko-Honokōhau was established in 1978 "in order to provide a center for the preservation, interpretation, and perpetuation of traditional native Hawaiian activities and culture, and to demonstrate historic land use patterns as well as to provide a needed resource for the education, enjoyment, and appreciation of such traditional native Hawaiian activities and culture by local residents and visitors..." (National Park Service, Kaloko-Honokōhau, 2013). The area was first nominated and listed on the National Register as the Honokōhau Settlement in 1966. The Honokōhau Settlement shows the close relationship between the early Hawaiians and their environment. The area contains numerous Native Hawaiian archaeological and historic materials including three fishponds, several heiaus (temples), at least 50 residential house sites, and places of mortuary significance (Cordy et al., 1991). Today it is co-managed by the U.S. National Park Service and a Native Hawaiian descendant group of people affiliated with Kaloko-Honokōhau, and decision-making is done by a joint advisory group (Alous, 2015). Scientific research in Kaloko-Honokōhau is also coordinated with the joint advisory group.

National Monuments in the Southwestern United States

When the U.S. National Park Service was established in 1916, efforts were made to set aside additional protected areas, which included national monuments and wilderness areas. These areas included, in some cases, a sizable number of archaeological sites and places of significance to the tribes that had occupied them. While the monuments did not see removals of their Indian residents initially upon their establishment, some places, such as Chaco Canyon, New Mexico, saw the few remaining Diné (Navajo) removed by 1947 (Davina Two-Bears, personal communication, August 2021). Table 2.2 presents data on the national monuments and historical parks that were established in the Southwestern United States between 1907 and 2016.

All of these monuments and historical parks contain Ancestral Puebloan materials. They also contain shrines that the current affiliated tribes wish to have access to. The Pueblo community of Cochiti, for example, helps to co-manage Kasha Katuwe Tent Rocks National Monument which is on ancestral Cochiti land. The Navajo (Diné) have co-management rights over Canyon De Chelly National

Table 2.2 National Monuments and Historical Parks in the Southwestern United States and Their Affiliated Tribes

Name and state	Date of establishment	Size (km²)	Number of affiliated tribes
Bandelier National Monument, New Mexico	1916	136.29 km²	7
Canyon de Chelly National Monument, Arizona	1931	339.5 km²	2
Bears Ears National Monument, Utah	2016	1,351,849 acres, 5470.74 km²	5
Chaco Culture National Historical Park, New Mexico	1907 (redesignated as a National Historical Park (1980)	137.5 km²	25
Mesa Verde National Park, Colorado	1906 (29 June)	52,485 acres (21,240 ha), 212.4 km²	26
Totals	5	6296.43 km²	

Note: Data obtained from the National Park Service, U.S. Department of the Interior, Washington D.C. and from the websites of the individual monuments and national parks and specific tribes

Monument. In the latter case, there are at least 40 families continuing to reside in the national monument. In many ways, Canyon de Chelly stands as a potential model for other national monuments and protected areas where co-management provides multiple benefits both to native people and visitors and to park authorities.

Tourism and the National Parks

One of the arguments made by U.S. National Park Service personnel for the value of national parks is that they not only protect landscapes, habitats, and species, but they also serve as places where tourists (recreational visitors) can visit and enjoy 'the great outdoors.' Data on tourist visits to three national parks in the United States are presented here:

Yellowstone National Park: 8.991 km² (3472 mi²).
Established on March 1, 1872
Yellowstone December 2018

Concessionaires:

About 3200 people work for concessioners in the park at summer peak

National Park Service employees:

Permanent 382 total (177 full time year-round, 202 career seasonal, 3 part time)
Seasonal 370

Visitors

2021	4,860,537
2020	3,806,306
2019	4,020,287
2018	4,114,999
2017	4,116,524
2016	4,257,177

Yosemite National Park 747,956 acres (1168.681 sq. mi; 302,687 ha; 3026.87 km^2) (1169 mi^2).
Established on October 1, 1890
National Park Service employees 800 summer, 478 winter
Yosemite hospitality and concessions 1300 summer 800 winter

Visitors:

2021	3,345,988
2020	2,360,818
2019	4,586,463
2018	4,161,087
2017	4,479,242
2016	5,217,114

Glacier National Park 1,013,322 acres (4100.77 km^2) (1583 mi^2)
Established as a national park on May 11, 1910
Seasonal workers: around 350 in summer. Much of park in not open in winter and no employee numbers are listed

Visitors:

2021	3,080,000
2020	1,698,864
2019	3,049,839
2018	2,965,309
2017	3,305,512

In 2020, the numbers of tourists visiting the national parks in the United States declined substantially because of the coronavirus pandemic. By 2021, this trend was reversed. During the pandemic, there were cutbacks in the numbers of personnel working in the parks, including federal employees, individuals working for concessionaires, and volunteers. Incomes of people living on the peripheries of the national parks who had businesses also declined but picked back up in the summer of 2021. The impacts of the COVID-19 pandemic on the wild animals in the national parks are currently being assessed, but it is likely that at least some of them are on the increase, as indicated by the National Park Service's decision in December 2020 to cull 700 bison in Yellowstone. Culling, a controversial topic, is aimed at bringing the population more in line with levels that are considered sustainable in national parks (Gordon, 2019). It is interesting to note that the National Park Service has okayed the killing of a dozen buffalo in Grand Canyon National Park in 2021, allocating the hunting rights to private hunters who applied for the licenses to dispatch the animals through a lottery system (Scully, 2021).

Indigenous people, for their part, would like to be given the same chance as professional hunters to engage in culling activities and to utilize the meat and the hides, something that took place in southern Montana in March 2023 (Douglas McDonald, personal communication, 31 March 2023). At the same time, many indigenous people feel that the US national parks should be off-limits to recreational hunters. Establishing a set of regulations regarding extractive activities in national parks that satisfies both the National Park Service and associated tribes is clearly something that would go a long way towarding satisfying all of their stakeholders.

Conclusions

The declaration of national parks and monuments and other kinds of protected areas has had substantial impacts on indigenous people. The most common arguments made by governments and wildlife biologists have been that indigenous peoples tend to overexploit the wildlife and other natural resources in the areas where they reside. There are also arguments that indigenous peoples and their domestic animals are a threat to wildlife. Indigenous peoples and their supporters, on the other hand, have argued that they are, as Alec Campbell (1977: 40–41) has put it 'conservationists par excellence.' While indigenous peoples are sometimes blamed for poaching and for the over-exploitation of resources, there is mounting evidence that this is not generally the case.

Indigenous peoples employ numerous traditional and contemporary natural resource management strategies, including diversification of their exploitation of resources as numbers decline, purposeful management of resources using fire and transplantation of desirable plants, and engineering of the natural landscape in various ways in order to prevent over-exploitation of resources. Indigenous peoples continue to press for rights of access to national parks and other protected areas. They, like ecologists (e.g. Balmford et al., 2002) see both social and economic reasons for conserving wild nature.

References

Alous, R.-R. T.L. (2015). *Reauthorizing kanaka 'Oiwi heritage discourse at Kaloko-Honokohau National Historic Park*. M.A. thesis, Simon Fraser University.

Anderson, G. C. (2014). *Ethnic cleansing and the American Indian: The crime that should haunt America*. University of Oklahoma Press.

Balmford, A., Bruner, A., Cooper, P., Costanza, R., Farber, S., Green, R. E., Jenkins, M., Jefferiss, P., Jessamy, V., Madden, J., Munro, K., Myers, N., Naeem, S., Paavola, J., Rayment, M.,

Rosendo, S., Roughgarden, J., Trumper, K., & Kerry Turner, R. (2002). Economic reasons for conserving wild nature. *Science, 297*, 950–953.

Bangs, E. E., et al. (2008). Managing wolf–human conflict in the northwestern United States. In R. Woodroffe, S. Thirgood, & A. Rabinowittz (Eds.), *People and wildlife conflict or coexistence?* (pp. 340–356). Cambridge University Press.

Beschta, R. L., Ripple, W. J., Boone Kauffman, J., & Painter, L. E. (2020). Bison limit ecosystem recovery in northern Yellowstone. *Food Webs, 23*.

Black, G. (2012). *Empire of shadows: The epic story of Yellowstone*. St. Martin's Press.

Brockington, D. (2002). *Fortress conservation: The preservation of the Mkomazi game reserve, Tanzania*. James Currey.

Burnham, P. (2000). *Indian country, God's country: Native Americans and the National Parks*. Island Press.

Campbell, A. C. (1977). Conservationist par excellence. *Botswana Magazine, 2*(3), 40–45.

Callenbach, E. (1995). *Bring back the buffalo: A sustainable future for America's Great Plains*. Island Press.

Cannon, K. P. (1993). Paleoindian use of obsidian in the Greater Yellowstone area: New evidence of the mobility of early Yellowstone people. *Yellowstone Science*, 6–9.

Cordy, R., Tainter, J., Renger, R., & Hitchcock, R. (1991). *An Ahupua'a Study: The 1971 Archaeological Work at Kaloko Ahupua'a, North Kona, Hawai'i*. Western Archaeological and Conservation Center Publications in Anthropology, No. 58. National Park Service, U.-S. Department of the Interior.

Cannon, K. P., Ryan, E., & Martin, H. (2023). *The application of strontium isotopes tracking Holocene bison in the greater Yellowstone ecosystem*. Paper presented in a symposium titled 'A Further Discussion of the Role of Archaeology in Resource and Public Land Management, Ken and Molly Cannon, organizers, 88th annual meetings of the Society for American Archaeology, Portland, Oregon, 28 March-2 April, 2023.

Chacon, R. J., & Mendoza, R. G. (2012). Discussion and conclusions. In R. J. Chacon & R. G. Mendoza (Eds.), *The ethics of anthropology and Amerindian research: Reporting on environmental degradation and warfare* (pp. 451–503). Springer.

Craig, D. R., Yung, L., & Borrie, W. T. (2012). "Blackfeet belong to the mountains": Hope, loss, and blackfeet claims to glacier National Park, Montana. *Conservation and Society, 10*(3), 232–242.

Duffield, J., Patterson, D., & Neher, C. J. (2006). *Wolves and people in Yellowstone: Impacts on the regional economy*. University of Montana.

Farrell, J. (2015). *The battle for Yellowstone: Morality and the sacred roots of environmental conflict*. Princeton University Press.

Geist, V. (1997). *Buffalo nation: History and legend of the North American Bison*. Voyageur Press.

Gordon, I. J. (2019). Adopting a utilitarian approach to culling wild animals for conservation in National Parks. *Conservation Science and Practice, 1*, e105. https://doi.org/10.1111/csp2.105

Greene, L. W. (1987). *Yosemite: The Park and its resources*. U.S. Department of the Interior, National Park Service.

Grindle, D. (2023). *Elk in the rockies: Interweaving the ethnographic present and the archaeological past toward more thoughtful animal management*. Paper presented in a symposium titled 'A Further Discussion of the Role of Archaeology in Resource and Public Land Management, Ken and Molly Cannon, organizers, 88th annual meetings of the Society for American Archaeology, Portland, Oregon, 28 March-2 April, 2023.

Isenberg, A. (2000). *The destruction of the Bison: An environmental history, 1750–1920*. Cambridge University Press.

Janetski, J. C. (1987). *The Indians of Yellowstone National Park*. University of Utah Press.

Kay, C. (2007). Were native peoples keystone predators? A continuous-time analysis of wildlife observations made by Lewis and Clark in 1804-1806. *The Canadian Field-Naturalist, 121*(1), 1–16.

Leshy, J. D. (2021). *Our common ground: A history of America's public lands.* Yale University Press.

Loendorf, L. L., & Stone, N. M. (2006). *Mountain Spirit: The sheep eater Indians of Yellowstone.* University of Utah Press.

MacDonald, D. H. (2018). *Before Yellowstone: Native American archaeology in the National Park.* University of Washington Press.

Madley, B. (2016). *An American genocide: The United States and the California Indian catastrophe, 1846–1873.* Yale University Press.

Meagher, M. M. (1973). *The bison of Yellowstone.* National Park Service.

Nabokov, P., & Loendorf, L. (2004). *Restoring a presence: American Indians and Yellowstone National Park.* University of Oklahoma Press.

National Park Service, Kaloko-Honokohau. (2013). *Kaloko-Honokohau Cultural Center: Environmental Assessment (Draft).* National Park Service, Department of the Interior.

National Park Service, Yosemite National Park. (2019). *Voices of the people: The traditionally associated tribes of Yosemite National Park.* National Park Service.

Quammen, D. (2018). *Yellowstone: A Journey though America's Wild Heart.* National Geographic Society.

Runte, A. (1990). Joseph Grinnell and Yosemite: Rediscovering the legacy of a California conservationist. *California History, 69*(2), 170–181.

Ruth, T. K., Buotte, P. C., & Hornocker, M. G. (2019). *Yellowstone cougars: Ecology before and during wolf restoration.* University Press of Colorado.

Scully, M. (2021). Don't let hunters kill bison inside a National Park. *New York Times,* 6 September 2021.

Sholar, C. (2004). Glacier National Park and the Blackfoot Nation's reserved rights: Does a valid tribal co-management authority exist? *American Indian Law Review, 29,* 151–172.

Smith, D. W., & Ferguson, G. (2012). *Decade of the wolf: Returning the wild to Yellowstone.* Globe Pequot Press.

Smith, D. W., & Peterson, R. O. (2021). Intended and unintended consequences of wolf restoration to Yellowstone and isle Royale National Parks. *Conservation Science and Practice, 2021*(3), e413. https://doi.org/10.1111/csp2.413

Smith, W. C., & Warrior, R. A. (1997). *Like a hurricane: The Indian movement from Alcatraz to wounded knee* (2nd ed.). The New Press.

Smith, D. W., Stahler, D. R., MacNulty, & Whittlesey, L. H. (2020a). Historical and ecological context for wolf recovery. In D. W. Smith, D. R. Stahler, & D. R. MacNulty (Eds.), *Yellowstone wolves: Science and discovery in the world's first national park* (pp. 3–12). University of Chicago Press.

Smith, D. W., Stahler, D. R., & MacNulty, D. R. (Eds.). (2020b). *Yellowstone wolves: Science and discovery in the world's first national park.* University of Chicago Press.

Spence, M. D. (1999). *Dispossessing the Wilderness: Indian removal and the making of the National Parks.* Oxford University Press.

Soave, T. (2018). Protecting the gateway to Yellowstone: The upper Yellowstone River and its options for survival. *Colorado Natural Resources, Energy, and Environmental Law Journal, 29*(1), 165–194.

Spence, M. (1996a). Dispossessing the wilderness: Yosemite Indians and the National Park Ideal, 1864-1930. *Pacific Historical Review, 65*(1), 27–59.

Spence, M. D. (1996b). Crown of the continent, backbone of the world: The American wilderness ideal and Blackfeet exclusion from glacier National Park. *Environmental History, 1*(1), 29–49.

Spence, M. D. (2000). *Dispossessing the wilderness: Indian removal and the making of the National Parks.* Oxford University Press.

Stark, K. J., Bernhardt, A. L., Mills, M., & Robison, J. A. (2022). Re-indigenizing Yellowstone. *Wyoming Law Review, 22*(2), 387–487.

Treuer, D. (2019). *The heartbeat of wounded knee: Native American from 1890 to the present.* Riverhead Books.

Treuer, D. (2021). Return the National Parks to the tribes: The jewels of America's landscape should belong to America's original peoples. *The Atlantic,* 20 May 2021.

Utley, R. (1984). *The Indian frontier in the American west, 1846–1890.* University of New Mexico Press.

Chapter 3
Coercive Conservation: Removals of Indigenous Peoples from Protected Areas in Southern Africa

Introduction

This chapter considers examples of removals of peoples from protected areas in southern Africa. Nearly all of these protected areas have witnessed coercive conservation strategies, where milaristic means were employed by nation-states or agencies to require indigenous people to relocate to areas outside of the park or monument boundaries. Non-government organizations. The relocations, which were carried out in Botswana, Namibia, South Africa, and Zimbabwe, among other countries, were done in such a way that the people affected were worse off after they were moved than they were previously. In some cases, the people relocated were hunter-gatherers, and in other cases they were agropastorists or farmers. One of the reasons given for the relocation was that the indigenous people were seen as a threat to the wildlife and other natural resources of the protected areas and that they were engaging in practices that were deemed deleterious to the parks and reserves, such as setting fires. Later on, the importance of tourism in protected areas was stressed by governments and park administrators.

Hundreds and up to several thousand indigenous peoples lost their rights to protected areas in the poast century and a half in southern Africa. Conservation was seen as a justification for the removal of indigenous people in a number of cases in southern Africa, as shown in Table 3.1.

In the cases of what became Kruger National Park, formerly a game reserve when it was set aside in 1898, there were people residing inside the boundaries including the Makuleke and Tonga (Carruthers, 1995). Park rangers were employed to prevent local people from exploiting the wildlife, including elephants, rhinoceros, lion, buffalo, and leopard. Patrols were carried out in the protected area, and people who were found to have killed a protected animals were arrested and jailed. One of the impacts of such a strategy was that families lost members who were crucial to their livelihoods. Similar situations occurred in other parks and reserves in southern Africa.

© The Author(s) 2023
M. Sapignoli, R. K. Hitchcock, *Anthropology and Ethics*,
https://doi.org/10.1007/978-3-031-39268-9_3

Table 3.1 National Parks, Game Reserves, and Conservation Areas in Southern Africa that resulted in the Involuntary Resettlement of local populations as a means of protecting habitats and wildlife

Park or Reserve Area, Establishment Date, Size in km^2	Country	Comments
Central Kalahari Game Reserve (1961), 52,730 km^2	Botswana	Over 2200 Glui, Gllana, and Boolongwe Bakgalagadi were resettled outside the reserve in 1997 and 2002
Chobe National Park (1961), 9980 km^2	Botswana	Hundreds of Subiya were resettled in the Chobe Enclave, where 5 villages are in a 3060 km^2 area
Etosha National Park (1907) Game Reserve No. 2, (93,240 km^2) reduced in size in 1970 to 22,912 km^2	Namibia	Haom San were resettled outside of the park and sent to freehold farms in 1954; new relocation on-going (2009-present)
Gemsbok National Park (1931), made transfrontier park in April, 1999, 37,991 km^2 now Kgalagadi Transfrontier Park	South Africa, Botswana	≠Khomani and Nlamani San were resettled out of the park in the 1930s; won co-management rights to the park in 1998
Wankie Game Reserve (1927), declared Hwange National Park, January 29, 1950 (14,651 km^2)	Zimbabwe	Tshwa San were rounded up and resettled south of Wankie Game Reserve in the late 1920s and early 1930s
Khaudum National Park (2007) (formerly Khaudum Game Reserve) (1989) 3841 km^2	Namibia	Jul'hoansi resettled in 1989 and 2007, mostly to Nhoma and //Ao//oba in the northern part of Nyae Nyae in Otjozomdjupa
Kruger National Park (1926), 19,485 km^2 now part of Greater Kruger National Park (GKNP)	South Africa	Protracted efforts to resettle Makuleke from Kruger culminated in the relocation of 1500 people in 1969
Moremi Game Reserve (1964), 3880 km^2	Botswana	Bugakwe (llAni-kxoe) San were relocated out of the reserve in the 1960s and 1970s
West Caprivi Game Park (1963), became Bwabwata National Park (2007) (6724 km^2) (2422 mi^2)	Namibia	Kxoe and Mbukushu were resettled out of the game reserve in the early 1960s, some resettled in 2017–2018

Case 1: Wanke Game Reserve and Hwange National Park, Zimbabwe

One of the first places in southern Africa where communities were relocated involuntarily from a protected area was Wankie Game Reserve, now Hwange National Park in what is now Zimbabwe. This relocation was a result of decisions by the then Southern Rhodesian administration to create a game reserve in the western part of the country bordering on the Bechuanaland Protectorate in 1927. The plan was to appoint a warden to oversee the game reseve, part of whose job was remove any people who were living inside of its boundaries. The relocation focused mainly on 'Bushmen', many of whom now self-identify as Tshwa San.

The Tshwa are a San-speaking group in western Zimbabwe and northern Botswana (Rankomise, 2015; Hitchcock et al., 2016; Ndlovu et al., 2022). They are former foragers who now are agropastoralists and wage laborers who engage in foraging primarily as a source of food or, to a lesser extent, income. Tshwa are found in western Zimbabwe, where they reside primarily in Tsholotsho District in Matabeleland North Province and Bulilima Mangwe District in Matabeleland South. In Botswana Tshwa are found in the Makgadikgadi Pans, along the Nata River, and in the east-central Kalahari.

The Tshwa speak a 'Central San' (Khoe) language similar to the G/ui and G//ana of the western and central Kalahari and the Naro San of the Ghanzi Ridge (Güldemann, 2008, 2014; Pratchett, 2020, 2021). The Tshwa are sub-divided into a number of different named groups, including /Aise, Ganade, and Danisan (Hitchcock, 1982; Hitchcock et al., 2016). Tshwa and Shua differ somewhat from other peoples of hunting and gathering origin in southern Africa in that they were generally sedentary for a substantial portion of the year and were heavily integrated into the regional economies of Tswana, Ndebele, Kalanga, and other non-foraging southern African populations by the seventeenth, eighteenth, and nineteenth centuries.

The Tshwa were affected heavily by the creation of the Wankie Game Reserve, now the Hwange National Park, in the late 1920s. Wankie Game Reserve was originally declared in 1927 and it was established as Hwange National Park on January 29, 1950. The park is 14,651 km^2 in size and is the largest national park in Zimbabwe.

In the park itself, in the nineteenth century, Tshwa were seasonally mobile, and they obtained water through digging pits close to pans or in the beds of seasonal rivers. There was some use of Wankie by groups other than Tshwa, including Nambiya, Ndebele, and Kalanga, most of whom entered the area briefly for purposes of hunting. Some of the reasons that the government was reluctant to have native reserves of farmers created inside the Wankie area were (1) the soil fertility was low, (2) water distribution was problematic, and (3) there were patches of a plant that was poisonous to cattle, known *m'khauzaan* (or *mogau*, *Dichapetalum cymosum*) (Haynes, n.d.: 114). It is interesting to note that Haynes also says that people who were relegated to the native reserves on the boundaries of Wankie thought of them as 'cemetires, not homes' (Haynes, n.d.: 114). One reason for these beliefs was that both of the native reserves, and Wankie itself, were known to have tsetse fly (*Glossina morsitans*) which carried sleeping sickness and mortality rates from malaria were high.

The rinderpest epidemic of 1896–97, combined with extensive hunting of large mammals by both Europeans and local people, led to a reduction in wildlife numbers in Southern Rhodesia and many of parts of southern Africa (Van Onselen, 1972; Mutowo, 2001). Elephant populations in particular were disturbed considerably by hunters in the latter part of the nineteenth century. According to local people, one response of the elephants in the Wankie and Tsholotsho regions was a tendency to bunch up in small mixed herds. Without the leadership of the matriarchs, there was greater destruction of crops of local people, according to local community members. Some of the local farmers responded accordingly, opting to shoot the animals on sight, especially if they were in their fields. Elephants were considered problem animals both by colonial government authorities and by local people.

The depletion of wildlife fueled concerns in the British South Africa Company and the colonial government of Southern Rhodesia that the resource potential of the region would be lost unless steps were taken to stop the killing. One way to deal with the problem, it was decided, was to utilize the 'royal game' principle of the Ndebele and Kalanga chiefs and to declare wildlife species as state property. It was made illegal for individuals to kill game even if it invaded their fields or threatened their lives. As one Tshwa put it, "The Europeans became the gamekeepers, and the Africans became the poachers." Such actions of problem animal control were disallowed by the British South Africa Company and the Southern Rhodesian administration in the early part of the twentieth century.

In the period between 1890 and 1923, the Department of Agriculture oversaw the administration of game in Southern Rhodesia. The first full-time officer with responsibility for overseeing game management in Hwange, Ted Davison, was appointed in 1928. Davison, undertook trips into the Wankie area to assess its status and to tell Bushmen and other residents that they were breaking the law for continuing to live in Wankie (Davison, 1977: 17–24). These efforts were not easy, as noted by Davison, who said, "Bushmen who knew the area kept their secrets, refusing to divulge any information at all—probably because they felt this might lead to the arrest of relatives engaged in poaching" (Davison, 1977: 16). One his tasks, according to Davison, was to warn people that the area was now a game reserve and that they were not allowed to live there (Davison, 1977: 20).

Davison, unlike some other Southern Rhodesian wildlife personnel, had a certain amount of empathy for Bushmen. This is revealed in a statement he made in his book:

> These Bushmen, in fact, evoked a degree of sympathy. They were not really poachers in the worst sense. Just like a pride of lions, they killed only for their own needs, amounting to not much more than an animal a week. However, the law had come to Wankie Game Reserve and it had to be implemented (Davison, 1977: 21).

Unfortunately, there were other, less positively inclined individuals, some of whom worked for the Southern Rhodesian government, and others who were "self-appointed conservationists." One of these men, H.G. Robins, was a former hunter who resided on a farm to the north of Wankie Game Reserve. According to Davison (1977: 23), Robins was obsessed with the idea that the region was "infested with poachers, all of whom were concentrating their efforts on his land." Robins carried out patrols both by vehicle and on foot, looking for the tracks of Bushmen who he believed were responsible for what he saw as declining numbers of large game. Tshwa in the region described how Robins hunted people down and either beat them or turned them over to government authorities. He was also known for using 'shoot to kill' antipoaching strategies.

Davison concluded after some of his initial surveys of the Wankie region that the poaching problem was not nearly as serious as he had been led to believe (Davison, 1977: 23–24). He admitted that there were indeed Bushmen families moving around the area, some of them with muzzleloaders (Davison, 1977: 24). These Bushmen apparently were not using either poisoned arrows or wire snares, items that were considered by park rangers to be highly lethal to wildlife populations. In Davison's opinion, the biggest constraint affecting wildlife populations in Wankie was not poaching but rather the availability of surface water.

In the late 1920s the Tshwa were informed that they had to move out of the Wankie Game Reserve. Some of them did so, but others retreated into the dry interior of the game reserve along the Botswana-Zimbabwe border. Patrols were sent in to arrest people and to remove them from the game reserve. Tshwa were arrested and jailed, while their families attempted to eke out an existence in areas south of the park in what are now parts of Tsholotsho District. A few Tshwa went to areas north of the Wankie Game Reserve, and some moved to the Gwaai and Lupane areas where they worked for African farmers. Oral history testimony underscores the privation and hunger that occurred among the Tshwa who were resettled and prevented from engaging in hunting and gathering activities. As one Tshwa elder put it, "We were so hungry that we were forced to eat leaves and bark of trees" (Tshwa informant, December 1980).

According to ethnohistoric information and contemporary ethnographic interviews, the livelihoods of Tshwa declined significantly as a result of the move. Some of the Tshwa chose to enter into the local and regional economies, working in the coal mines of Hwange, at copper and chromite mines in Matabeleland South, and on commercial farms owned by Europeans. Others opted to remain as foragers, engaging in wildlife utilization and wild plant collection. About a dozen Tshwa joined the Zimbabwe National Parks and Wildlife Management Service, some of them working in Wankie Game Reserve and later, Hwange National Park.

Incidents in Northern Botswana and Transformations in Western Zimbabwe

Unfortunate events that occurred in the 1943–45 period saw a group of Ganade Tshwa from Gumg//abi in Bechuanaland accussed of murder of two Royal Air Force Cadet pilots from Kumalo in Rhodesia, who had landed their airplane in Kaucaca Pan in October 1943 and disappeared. After investigations, the Bechuanaland Protectorate Administration put 8 Tshwa on trial at the High Court in Lobatse, but the men and women were acquitted on the basis of lack of evidence (Laverick, 2015; Hitchcock et al., 2016; Skidmore-Hess, 2021). The Bechuanaland Protectorate Administration and the Bamangwato Tribe sent groups of men, some of them mounted on camels, into the northern Makgadikgadi area and forced possibly as many as 1200 Tshwa to relocate to places south of the Nata River in Botswana where they were required to remain in settled villages. The Tshwa had their weapons cofiscated, and some of them ended up working for the Ndebele, Kalanga, and Bamangwato who occupied the border region. A few Tshwa returned to the Northern Crown Lands (later Northern State Lands) in small groups for hunting and gathering purposes, traveling at night in order to avoid detection.

Groups of Tshwa who resided in the Wankie area utilized the seasonal pans and rivers, wildlife, and vegetation before their eviction and resettlement in 1927–1930. The majority of Tshwa in Southern Rhodesia were moved to 'native reserves' which are communal land. The land arrangements, in line with Zimbabwe government policy, is that all communal land is state land (Scoones et al., 2011). The Tshwa, therefore, did not have *de jure* (that is, legal) rights to the land that they occupied. They were able to obtain plots of land for residential, agricultural, and income generation purposes from traditional authorities or the central government, but they potentially could lose their land at any time, something that several informants noted had happened to them in the the 1920s, 1930s, and 1940s as well as recently.

Oral history data indicate that the Tshwa who were relocated out of the Wankie Game Reserve and the Northern Crown Lands of Botswana shifted toward a more mixed economic system in which working for livestock producers played a significant role. Some of the Tshwa male household heads became herders (*badisa*) for Ndebele, Kalanga, and Ngwato cattle owners, receiving milk, grain, and sometimes a cow a year in exchange for their labor (Hitchcock & Nangati, 1993; Hitchcock et al., 2016, 2020). Some of the Tshwa households raised crops, especially sorghum, millet, melons, millet and, by the mid-twentieth century, maize and beans. An important source of food and income for many Tshwa was mopane worms (*Imbrasia belina*). Having access to the park for wild insect and timber resources were key requests of Tshwa which they made to Southern Rhodesian officials and later Zimbabwe government officials (Hitchcock et al., 2016).

In the 1980s and 1990s, the government of Zimbabwe embarked on programs aimed at promoting, at least to a limited extent, community control of wildlife (Child, 1995; Dzangerai, 1995; Mberengwa, 2000). CAMPFIRE—the Communal Areas Management Programme for Indigenous Resources—was established in 1986, and over the next two decades had significant impacts on district councils and, to a lesser extent, rural communities. The rules of the CAMPFIRE Programme were such that funds went to district councils, such as Tsholotsho, rather than the dozens of communities in the District.

The establishment of national parks in Zimbabwe has led to the dispossession of resident populations, as occurred not only in Hwange but also the Matopos, Zambezi National Park, and Gonarezhou (see Table 3.2). Generally, the evictions occurred at the time or slightly before the establishment of the protected area, as seen, for example, in Hwange and Gonarezhou. In few, if any, cases, was compensation paid or alternative land made available to the resettled people who were considered 'oustees'.

One of the major concerns of Tshwa today in Tsholotsho is the fear of being resettled again. Approximately 10% of the population of Tsholotsho, Zimbabwe, and the Nata River region of Botswana said in interviews that their ancestors had

Table 3.2 National Parks and Game Reserves in Zimbabwe

National Park or Game Reserve	Date of founding	Size (in sq km)	Events occurring there
Hwange (Wankie) National Park	1927—Game reserve, January 29, 1950 national park	14,651 km^2 5652 mi^2	Tshwa San evicted 1927–1930
Matusadona National Park	1963—game reserve 1975—national park	1407 km^2 543 mi^2	Local communities evicted
Matobo (Matopos) National Park	1926—monument 1930—national park	425 km^2 164 mi^2	Some Ndebele and Kalanga resettled
Nyanga National Park	1989	314 km^2 121 mi^2	
Chizira National Park	1990	1920 km^2 741 mi^2	
Mana Pools National Park (Lake Kariba) Urungwe District, Mashonaland North	1989	2196 km^2 848 mi^2	With Sapi and Chewore Safari areas, WHS 6766 km^2
Zambezi National Park	1989	570 km^2 220 mi^2	Communities relocated
Gonarezhou National Park	1934—Game sanctuary, 1975—National park	5053 km^2 2930 mi^2	Shangaan and other groups evicted 1957–68
Chizira National Park	1972	1920 km^2 741 mi^2	
Victoria Falls National Park	1989	23 km^2	2340 ha in Zimbabwe, WHS
Zambezi National Park	1989	7.41 km^2 Along river	741 ha riparian zone WHS
Mosi-oa-tunya National Park	1989	66 km^2 mi^2	3779 ha (Zambia) WHS
Great Zimbabwe National Monument	1972	722 ha 7.22 km^2	World Heritage Site
Khami Ruins National Monument	1937	108 ha 1.08 km^2	World Heritage Site
Totals	14 protected areas	201,037.3 km^2	Several are transboundary areas

Note: Data obtained from Zimbabwe National Parks and Wildlife Management Authority (ZNPWMA); Zimbabwe Ministry of Environment, Water, and Climate, (ZMEWT); WHS stands for World Heritage Site

been resettled as a result of the establishment of the Wankie Game Reserve or by the establishment of native reserves in the Wankie area and commercial farms belonging to the Colonial Development Corporation (CDC) in the northern Makgadikgadi Pans region of Botswana (Hitchcock & Nangati, 1992, 1993, 2000; Hitchcock et al., 2016, Hitchcock, field data, 1975–76, 1980–1981, 1992, 1995, 2012, 2013; Davy Ndlovu, field data 2020, 2022). Some resettlement of Tshwa took place during the liberation struggle (1965 to 1980) and some during the period of 'troubles' (known as 'Gukurahundi' in Shona, which is used to describe the spring rains that wash away the chaff from the wheat). This was a period when Zimbabwean government (ZANU-PF) forces sought to eliminate 'dissidents' in Matabeleland and the Midlands regions from 1981–1988 (Catholic Commission for Justice and Peace in Zimbabwe and Legal Resources Foundation, 2008; Ngwenya, 2018).

Resettlement pressures around Hwange increased substantially in September, 2013, when the carcasses of elephants and other animals were discovered in he southern portion of the park and in areas outside of the park in Tsholotsho. There were indications that cyanide was used to kill the animals (Muboko et al., 2014, 2016). Ivory was taken from some of the elephant carcasses, indicating potential poaching or scavenging. Subsequently, arrests were made of over two dozen people from Tsholotsho, Bulawayo, and other places for alleged involvement in the procurement, distribution and use of cyanide. None of these, it turned out, were Tshwa. After the elephant and other animal deaths were discovered, people residing in the areas close to the southern boundary of Hwange National Park were told by government and provincial officials that they had to move to new places away from the southern boundary of the park including some Tshwa and Ndebele families. However, they had not been informed of any relocation plans or compensatory measures as of March 2023.

Members of indigenous groups are often blamed for involvement in illegal hunting even though the evidence suggests that they play a minor role in poaching. Pressure on the San and other indigenous peoples continues in southern Africa, with arrests and sometimes shooting of people for suspected violation of hunting laws expanding, at least for a period of time, in Zimbabwe and Botswana. The numbers of people arrested for alleged poaching increased during the coronavirus pandemic, in spite of the fact that the numbers of wildlife department personnel have been reduced as a result of cost-saving measures. Resettlement of people away from the borders of Hwange National Park continues to be a rallying cry in Zimbabwe, but it is opposed vehemently by local people and the non-government organizations that support them, including the Tsoro-o-tso San Development Trust (TSDT). The most serious problems that continues to face the Tshwa and their neighbors today are poverty and insecurity of land tenure.

Case 2: The Central Kalahari Game Reserve Case, Botswana

The removals of people from the Central Kalahari Game Reserve in Botswana has likely been one of the most heavily documented case of involunary relocation out of a protected area in the late twentieth and early twenty-first centuries. These relocations occurred in 1997, 2002, and 2005, and they had significant impacts on the people who were resettled.

The Republic of Botswana, which gained independence on 30 September, 1966, has long been known as one of the world's most successful democracies. It has had eight open, multi-party elections, the most recent being in October 2019, and has had a strong record of economic growth. One area where Botswana has had problems, however, is with its relations with its minority populations, including the San and Bakgaladi of the Central Kalahari region.

Established in 1961 on the recommendation of a Bechuanaland Protectorate administrative officer and anthropologist, George Silberbauer, the boundaries of the reserve were gazetted formally in 1961 under the Protectorate's *Fauna Conservation Proclamation* (Bechuanaland Protectorate, 1961). As the Bushman Survey Officer for the Bechuanaland Protectorate, Silberbauer was fully aware that the boundaries were artificial, and that people recognized land outside the reserve as theirs for their use. What George Silberbauer proposed originally that the Central Kalahari become a 'people's reserve' which would be set aside for the protection of the lifestyles and livelihoods of the residents, the majority of whom at the time were G/ui and G//ana San. The Bechuanaland Administration, notably George Winstanley, disliked the idea of a 'people's reserve', citing some of the problems of native American reservations in the United States, so it was decided that the area become a game reserve instead. This designation as a game reserve resulted in restrictions being placed on the residents, particularly as regards hunting and the keeping of livestock (Bechuanaland Protectorate, 1963; Silberbauer, 1965, 1981, 2012; Hitchcock, 2002).

By the 1980s, however, the Botswana government and some ecological researchers were becoming increasingly concerned about the concentration of people around!Xade in the Central Kalahari, where there was permanent water and various social services including a school, health post, and other facilities.!Xade had over 1200 people living there by 1983 (Tanaka, 1987). After holding a CKGR Commission, the government opted to encourage people to move out of the reserve 'in order to facilitate their development' (Government of Botswana, 1985). As is stated on the website of the Owens Foundation (www.owensfoundtion.org, accessed 16 December 2022) Mark and Delia Owens, well-known conservationists who worked in the Central Kalahari from 1974–1981, made the following statement:

> During their seven years of working for conservation in the Kalahari Desert of Botswana, Delia and Mark Owens were greatly impressed with the Bushmen who live in this vast, harsh wilderness. Delia and Mark believe that the San belong to the Kalahari and should be allowed to continue their hunter-gatherer lifestyle in harmony with the natural balance of the Kalahari. However, should any members of the Bushmen decide to alter their lifestyle to include agriculture and the raising of livestock, it would be destructive to the fragile ecological balance of this protected wilderness area. Delia and Mark believe that Bushmen wishing to raise crops and livestock should be allowed to live on the lands surrounding the Central Kalahari Game Reserve where their activities will not endanger the survival of the plants, animals and people that currently live within the Reserve in harmony. Delia and Mark hope that the Bushmen will continue their amazing hunter-gatherer lineage as an inspiring part of the Kalahari ecosystem (Owens Foundation, www.owensfoundtion.org, accessed 16 December 2022).

It is important to note that it was conservationists who made the original argument that the residents of the reserve should be removed from the land that had been set aside for them under the Bechuanaland Protectorate colonial administration in 1961.

Alec Campbell, a deputy district commissioner under Mr. Silbebauer in 1962, said that there were G/ui, G//ana, Tsila and other San and Baboalongwe Bakgalagadi living in the reserve in the late 1950s and early 1960s (Alec Campbell, personal communication, 2011). He also pointed out that some of the people who lived periodically in Kikao and Gugamma actually had their major homes in Salajwe in the Kweneng District while people living in Molapo recognized rights to areas in Central District close to the Boteti River such as Rakops. As both he and Silberbauer noted, given the prevailing political conditions, there was no way in which the boundaries could have been gazetted to make the reserve larger, and there was no way that it could be declared a 'peoples' reserve' since there was no legislation in Bechuanaland that would have allowed for that at the time (Silberbauer, 2012).

There is significant sociocultural diversity in the Central Kalahari which had to be taken into consideration in the work and planning of the reserve (Tanaka, 1980, 2014; Tanaka & Sugawara, 2010; Silberbauer, 1981). The people who resided in the Central Kalahari included members of nine ethnic groups: G/ui, G//ana, G//olo, ‡Hoan, Kũa, Tsassi, Ts'aokohoe, and Tsila San and Babalaongwe Bakgalagadi. Members of some of these groups spoke only only mother tongue languages, so paying attention to translation was absolutely crucial. The two main Central Kalahari San groups were the G/ui and the G//ana. The Gui and G/ana, who are two distinct ethnic groups who have considerable rates of intermarriage, both speak what is known as a Khoe-Kwadi language, and they are part of the western Khoe sub-group of the Khoe-Kwadi (Güldemann, 2014: 27, Fig. 5).

At the time that Silberbauer and Campbell did the survey of the Bushmen in the Bechuanaland Protectorate as part of the Bechuanaland National Census in 1965, the G/ui and G//ana, along with the and Babalaongwe Bakgalagadi, represented the majority of the people living in the reserve. The total number of G/ui today is 1675, while the G//ana is 2285. The third largest group living in the CKGR is the Babalaongwe Bakgalagadi who collectively number over 5000 people, approximately 150 of whom were living in the Central Kalahari. The crucial factor facing all of these communities was access to potable water.

Although in the 1960s the Protectorate Administration retained control and regulation of the hunting and exploitation of wild animals in all parts of the Protectorate, it was left to the chiefs and headmen in the tribal areas to formulate and implement rules and regulations regarding wildlife usage (Hitchcock, 1988, 2000; Spinage, 1991). Certain animals (e.g. 'royal game' such as elephants and lions) were off-limits to hunters, as were animals seen as threatened or endangered because of their relatively small numbers such as springbok. In some cases, the hunting of specific animals was not allowed by order of Tswana chiefs, as was the case, for example, with giraffe among the Bangwaketse, Bakwena, and Bamangwato (Spinage, 1991: 10). As it turned out, the Bechuanaland Protectorate passed a series of laws to conserve wildlife and protect habitats in the country in the period between 1891 and 1966 (Spinage, 1991: 91–96), all of which affected the people of the Central Kalahari.

One of the few places where Bushmen and Bakgalagadi were allowed to hunt for subsistence purposes was the Central Kalahari Game Reserve. It should be noted, however, that there were disagreements over whether or not the residents of the Central Kalahari Game Reserve could hunt without restriction (see, for example, the *Regulations of the Central Kalahari Game Reserve*, Government Notice No. 38 of 1963). From 1979 to 2004, people in the Central Kalahari were allowed to hunt as long as they did so for subsistence purposes, had Special Game Licenses (SGLs), and providing that they used traditional weapons (bows, arrows, and spears) (Hitchcock & Masilo, 1995; Hitchcock, 2001). There was a significant debate over the issue of hunting rights in the 1980s and 1990s (Owens & Owens, 1981; Spinage, 1991). This debate, which involved government wildlife officials, ecologists, and development personnel, was fierce. In the end, the Botswana government opted to follow the recommendations of the ecologists and government personnel and recommended that people should be removed from the Central Kalahari and that all hunting in the reserve should cease.

In the mid-1980s, some Botswana government ministries argued that the residents of the Central Kalahari should be relocated to places outside of the CKGR. Several reasons were given. First, it was suggested that the people of the central Kalahari were no longer 'traditional' since they lived in stationary villages, kept domestic stock, and hunted with the aid of traps, horses, donkeys, and dogs (see, for example, Owens & Owens, 1981: 31). Second, it was argued that the people were having a negative impact on the wildlife in the reserve through their hunting activities. Third, it was noted that the reserve would have greater tourist value if it was 'pristine' (i.e. if there were no people living inside the reserve). Fourth, it was pointed out that having all of the people grouped in a single location outside of the reserve would facilitate the provision of development services by government, including water, health, and education. Finally, it was noted that if local people were relocated to other areas, it would be easier to deal with them administratively according to statements by the Government of Botswana (see Botswana National Archives BNA files).

In 1986, the government of Botswana announced that the Central Kalahari should be maintained as a reserve and that residents of the reserve should be encouraged to move elsewhere where they could take part in 'development.' A series of public meetings, known in Botswana as *kgotla* (public consultation) meetings, were held in which government officials discussed the findings of the commission and the decision of the government to have people resettle outside of the reserve. A complaint made by local people was that "Their voices were not heard" in these discussions because, as they noted, they had little opportunity to respond to the statements of government officials.

The years between 1988 and 1997 were taken up with periodic efforts by government officials to convince people to leave the reserve. There were also efforts on the part of the Department of Wildlife and National Parks and the Botswana Police to increase their anti-poaching operations in the CKGR, and dozens of people were arrested. In some cases, people suspected of peaching were beaten and otherwise mistreatred, resulting in an international oucry over Botswana's treatment of minorities (Mogwe, 1992). People from the Cental Kalahari, including the organization First People of the Kalahari, called for Botswana to allow people to remain in the reserve, as was done, for example, at the United Nations Human Rights meetings in Geneva in 1993 (Boustany, 1995). They also called for an end to Botswana's 'shoot to kill' policy (INK Center for Investigative Journalism, 2016; Mogomotsi & Madigele, 2017).

In spite of the international outcry, the first set of removals of the people from the Central Kalahari in 1997 saw 1239 people moved to New Xade and 500 people moved to Kaudwane in the northern Kweneng District (Ikeya, 2001: 188). After the relocation, the population of the CKGR was estimated to be between 420 and 450. In January, 2002, the Botswana government informed the remaining residents of the Central Kalahari Game Reserve that they were shutting down the wells and stopping all food deliveries inside the reserve. In February, 2002, the government began moving the people and their possessions out of the reserve (Sapignoli, 2018). Wells and boreholes were destroyed, livestock (goats, donkeys, and horses) were loaded on trucks or scattered in the bush, and people's homes and trees and other assets were burned. By March, 2002, it was estimated by former residents that there were less than two dozen people remaining in the reserve. The total number of people removed from the Central Kalahari was estimated to be approximately 2500.

This set of removals from the reserve was a classic example of involuntary relocation, or so that was what the residents of the reserve and some of their supporters argued. This relocation was done by government and district officals in such a way that local people were heavily impacted, with some residents of the reserve being chased down and thrown onto trucks where they were packed tightly together, and people who resisted were beaten and put on the trucks (Workman, 2009; Sapignoli, 2018). Families' homes were burned, their fruit trees uprooted, and their cooking pots and jerry cans for water were destroyed. The people were then removed from the reserve to places far away from where they lived originally. In some cases, individuals did not know where spouses, children, and other relatives were taken, and it took weeks or even months before they found their kin.

These issues became the subject of the legal case filed by Roy Sesana and Keiwa Setlhobogwa and 241 others against the Attorney General of Botswana (in his capacity as the recognized agent of the Government of the Republic of Botswana) (High Court of Botswana, 2002). The initial filing of the legal case was unsuccessful, and it was rejected on a technicality by the High Court in 2003. The case was appealed by the lawyers for the former residents of the reserve and it went to trial in July 2004; it lasted until December 2006. This trial, the longest and most expensive in Botswana history, has had important implications for the ways in which conservation-related resettlement has been approached since that time.

The verdict was read by the three judges on 13th December, 2006. The final judgment of the High Court was that (1) the government was not required to restore services in the reserve, (2) the stopping of services was lawful, and (3) the removals of people and denial of their land and subsistence rights in the Central Kalahari were unlawful, (4) that the removal was without the applicants' consent, (5) that all of the residents removed in 2002 had the legal right to return to the reserve and did not need a permit from the Department of Wildlife and National Parks to do so. Finally, the people of the CKGR were recognized as the indigenous peoples of the area. The problem was that the people of the CKGR theoretically have the same rights that any other citizens of Botswana have.

An important issue that arose out of the CKGR case and the efforts of the San and Bakgalagadi to promote themselves as indigenous peoples relates to the question of how to define "development." Development, from the perspective of the Botswana government, is seen as assistance that is aimed at enhancing the economy, living standards, and general well-being of the country's citizens (Saugestad, 2001). Indigenous peoples have a more holistic view of development, which revolves around the relationships of people, their environment, their ancestors, and their gods. The San and Bakgalagadi see the government's way of approaching development as being based on western principles of liberal capitalism. Instead, indigenous peoples' approach to development revolves around their relationship to "mother earth," to the land, and to one another and is more sustainable in terms of the environment, economy, and society.

Some of these arguments about what it means to be indigenous were heard from the lawyers of the applicants and the respondent (the government) during the two CKGR court cases of 2004–2006 and of 2010–2011. The issues surrounding the Central Kalahari Game Reserve are extremely complex (see Ng'ong'ola, 2007; Cook & Sarkin, 2009; Solway, 2009; Kiema, 2010; Sapignoli, 2009, 2015, 2016, 2018; Saugestad, 2011; Zips-Mairitsch, 2013; Morinville & Rodina, 2013). Issues about which there were debates ranged from the length of time people had occupied the Central Kalahari and the ways in which people survived in the area to the self-identification of people as indigenous (Sapignoli, 2018).

A major challenge for the people of the Central Kalahari has been the failure of the government of Botswana to implement the decisions that had been reached in the Central Kalahari Game Reserve legal case that concluded in December, 2006 (see the transcripts of the legal case and the judgment (High Court of Botswana Transcripts, 2006); see the analysess of the case by Solway, 2009 and Sapignoli, 2018). After they won the case, former residents of the reserve tried to enter the Central Kalahari in early 2007 but they were stopped from doing so by government officials because of the lack of entry permits. Later on, small groups of people quietly went in to the reserve where they made their living largely as hunters and gatherers, supplementing their subsistence with food they brought with them or which they purchased in the settlements on the peripheries of the reserve.

There are many different positions on the ways in which the Botswana government handled the resettlement and the legal cases arising from it. A sizable portion of the Botswana populace who have been interviewed by the media and by researchers have maintained that the Botswana government is violating the rights of the people in the Central Kalahari by refusing to allow them access to water, health services, medicines, development assistance, and food commodities that other people in Botswana get (Seleka et al., 2007; World Bank, 2015). They were also concerned about the fact that residents of the Central Kalahari continue to be arrested for violating conservation laws, in spite of the fact that the 2006 CKGR legal case reaffirmed their right to subsistence hunting.

From the standpoint of the NGOs supporting the people of the Central Kalahari, including Survival International, the Working Group of Indigenous Minorities in Southern Africa, the Forest Peoples' Programme, Land is Life, Minority Rights Group International, and the International Work Group for Indigenous Affairs, the government of Botswana is denying the human and development rights of the people of the Central Kalahari and, even more egregiously, denying them the right to water which they had won in the Botswana Court of Appeal in 2011 (see Botswana Court of Appeal, 2011; Sarkin & Cook, 2010–2011).

There were a number of changes that occurred as a result of the conservation-related resettlement in the late 1990s and early 2000s, including the following: The nutritional status of the resettled population declined. The population growth rates (fertility) increased from approximately 0.05% in 1960 to 3–4% in 2020. Morbidity (illness) rates increased with higher population densities in the gazetted settlements and with more contact among various groups. Mortality rates increased as a result allegedly of disease, nutritional, and psychological stress. It appears that cardiovascular diseases increased, as did diabetes, tuberculosis, hookworm, and sexually transmitted diseases (STDs). HIV/AIDS rates increased, especially in the late 1990s and the new millennium, as was seen, for example, in the resettlement sites of Kaudwane, New Xade, and Xere outside of the Central Kalahari Game Reserve. The government had stopped all mobile health visits to the Central Kalahari in 2002, and it was not until 2016 that decisions were made to begin government services in the reserve including the provision of water. As yet, however, additional boreholes have yet to be provided in the Central Kalahari.

Resettlement has led to increased poverty rates and social inequality in remote area settlements. Particular poverty problems are found among children and the elderly. It appears than larger households with more children have higher rates of poverty than do smaller households (field data, Sapignoli & Hitchcock, 2013). While the government attempted to offset the health problems in the resettlement sites by providing medical assistance; the problem was that often there were no medicines available to the clinics and health posts that existed outside of the CKGR in the resettlement sites. Health workers were not available, and crucial drugs such as malaria pills and anti-retrovirals (ARVs) were not provided in sufficient amounts to the health facilities according to local settlement residents. Health facilities were only available in the resettlement villages, but generally anti-retroviral medications and other medicines were unavailable.

The CKGR Residents Committee sent a letter to the central Botswana government in November 2018, requesting that the government not implement their plans for having the proposed Memoghamoga Community Trust (MCD), designed by a law firm, Lecha and Associates, go forward but instead allow each of the five communities in the CKGR to have their own individual community trusts. There was no response to this request as of March 2023. The communities in the CKGR continue to lack the rights to contol their own tourism or to benefit directly from tourism.

Table 3.3 presents data on the Central Kalahari communities including the resettlement sites. The three resettlement locations have seen an overall increase in

Table 3.3 Population and location data for communities in the Central Kalahari Game Reserve and the three resettlement sites

Name of community	Latitude and longitude	2014 population	2015 population	2019 population	2021–2022 population
Central Kalahari					
Gope (Ghagoo)	22°37′2.90″S 24°46′19.18″E	24	30	90	120
Gugamma	23° 6′55.34″S 24°15′27.47″E	0	29	28	0
Kikao	23° 1′42.21″S 24° 5′36.80″E	25	26	0 (10 on occasion)	0
Matswere	21° 9′24.21″S 24° 0′24.57″E	0	0 (DWNP staff only)	0	0
Menoatshe	22°41′2.94″S 23°58′33.13″E	0	0	Utilized for gathering	0
Metseamonong	22°25′12.59″S 24°13′02.76″E	120	130	46	56
Molapo	21°57′40.70″S 23°55′46.11″E	130	120	56	86
Mothomelo	22°06′39.19″S 25°01′59.61″E	150	26	77	91
!Xade	22°20′20″S 23°00′27″E	0	0 (DWNP staff only)	0	0
Xaxa	22°17′21.91″S 23°35′14.93″E	0	0	0	0
Total		449	362	317–330	353
CKGR resettlement sites					
New Xade	22°12′11″S 22°41′84″E	1269	Ghanzi District	1900	2100
Kaudwane	23°22′53.37″S 24°39′34.67″E	1084	Kweneng District	1700	1900
Xere	21°8′21.57″S 24°18′49.50″E	343	Central District	500	600
				4100	4600

Note: Data obtained from surveys and population censuses

the number of residents, a number of them consisting of non-San people coming in and taking land.

The years 2022 and early 2023 have seen increased pressure on the people of the Central Kalahari. The Ghanzi District Council failed to provide assistance in the form of COVID-19 related materials inside the CKGR. Vaccinations were provided only in the resettlement sites. Provision of food and water inside the CKGR continues to be sporadic. Community members are still complaining about human-wildlife conflict with lions and other predators, and they are worried about the expansion in the number of elephants that are destroying their gardens. The residents of the four extant communities inside the Central Kalahari continue to witness efforts made by the government of Botswana and their representatives to remove them.

The sociopolitical situation in the CKGR continues to be complex, particularly with the uncertainty surrounding the rights of residents inside the reserve, including restrictions on their rights to take part in tourism and to bury their dead inside of the reserve.

Case 3: Etosha National Park in Namibia

In 2001, James Suzman, who helped coordinate a Southern Africa-wide study of San peoples, made the following observation:

> For San in Namibia, land dispossession has been more extreme in both extent and form than for San elsewhere in southern Africa. The apportioning of the country under apartheid into freehold commercial farms, "tribal" communal lands and wildlife conservation areas meant that by 1976 fewer than 3% of the Namibian San population retained even limited *de jure* rights to the lands they had traditionally occupied. Close to half lived on freehold land owned by white farmers, for whom they worked and on whose employment they depended to retain basic residential rights (Suzman, 2001a: 11).

The indigenous population in South West Africa, now Namibia, that was dispossessed most extensively in the twentieth century was the Hai//om San. Today, the Hai//om are the largest San population in Namibia, numbering between 11,000 and 15,000. They are also some of the most widely distributed of the San people in the country. The Hai//om are found in the northern and north central parts of Namibia, stretching from the Oshikoto Region in the north south to Outjo and the commercial farms south of Etosha National Park, from Kunene Region on the west to Grootfontein in Otjozondjupa Region on the east. Hai//om are also found in central Namibia in the region that includes Otavi, Tsumeb, and Otjiwarango, with some Hai//om in the Grootfontien Farms and a few in N ≠ a Jaqna. Some Hai//om are found in the Waterberg area and in the commercial farms south of Otjiwarango.

The Hai//om are divided into a number of different sociolinguistic groups according to linguists, anthropologists, and to the Hai//om themselves (Hahn et al., 1928; Dieckmann, 2007: 112, Table 4.1). Some northern Hai//om use the term /Aekhoe as a term of self-identification (Widlok, 1999). There were 7506 Hai//om in the 1991 Namibia national census. More recent estimates suggest that there are at least 10,000–15,000 Hai//om who identify themselves as such in Namibia.

In the past, the Hai//om were hunter-gatherers who were involved heavily in exchanges of goods and services, trading high value products including ivory, ostrich feather and ostrich eggshell beads, copper items, grain, tobacco, honey, and calabashes for domestic foods, livestock, and household items (Gordon & Douglas, 2000: 26, 29). In the nineteenth century, based in part on control of exchange networks and on competition and conflicts among groups in Namibia for resources and power, relatively powerful leaders emerged among the Hai//om. Some of these leaders were instrumental in helping to maintain Hai//om autonomy in the face of encroachment of other groups.

The Hai//om today pursue mixed economic patterns, combining a small amount of foraging with agriculture, pastoralism, small-scale businesses, and wage labor. Sizable numbers of contemporary Hai//om reside on commercial farms and in the informal settlements in some of the towns of north-central and central Namibia such as Outjo and Otjivarango. A number of Hai//om also work in the mines of Namibia and various mines in South Africa.

In the rural areas, gathering of wild plants is done by Hai//om for purposes of food, medicine, and materials for building and other purposes. Traditional ecological knowledge of plants, animals and insects still exists, although some of these traditions are being lost as elders pass away and the young attend school. There are on-going efforts to preserve traditional knowledge and to document the Hai//om language. There have also been efforts to identify and map the traditional areas used by the Hai//om (see Vogelsang, 2005; Dieckmann, 2007). Based on data collected in 2002–2003, Hai//om elders identified the location of some 40 of their original settlements within Etosha National Park (Dieckmann, 2009).

The history of the Hai//om has been one where they experienced being removed from their ancestral lands through such processes as the creation of commercial farms, the enlisting of laborers for farm and other work, the establishment of colonial police posts (e.g. ones at Namutoni and Okaukuejo), and the declaration of the game reserves in the early part of the twentieth century (1907) (Dieckmann, 2003, 2007; Suzman, 2004). Most of the Hai//om who lost their lands ended up working on commercial farms while some were retained as trackers, scouts and laborers by the Department of Nature Conservation (DNC) in the Etosha Game Reserve.

Etosha originally was part of a large Game Reserve, called Game Reserve No. 2, which had been established by the German administrators of what was to become South West Africa in 1907 (Berry, 1997). The German administration allowed the Hai//om to continue to reside in and utilize wildlife and other resources in Game Reserve No. 2 until the end of the German colonial administration in 1915. It was the largest name reserve in South West Africa, (93,240 km^2) but later was reduced in size in 1970 to 22,912 km^2 in area. The Nature Conservation section of the South West African government decided to move the estimated 400 to 500 Hai//om-people living in Etosha to alternative places in the early 1950s.

In 1949, the South West African administration appointed at two-person Commission for the Preservation of the Bushmen. It was chaired by a former

Stellenbosch University professor, P.J. Schoeman, who was later to become the Chief Game Warden in Etosha. As Taylor (2009: 426) notes, this commission "had far-reaching effects on all Namibian San groups in terms of identity politics and land appropriation," a point echoed by Dieckmann (2007: 186). Schoeman, through his writings on Bushmen for example, *Jagters van die Woestylnland* (*Hunters of the Desert Land*) (Schoeman, 1951a), helped popularize stereotypes of the San as pristine hunter-gatherers and as people capable of surviving in marginal environments.

The commission produced an interim report in September, 1951 in which two Bushmen reserves were recommended: one for Khaung (!Kung) and another for the "Heikom" (Hai//om) (Schoeman, 1951b). When the final report came out in 1953, however (see Schoeman, 1953), there was only one Bushman reserve recommended, that of "Bushmanland" which was where the Ju/'hoansi lived (now Tsumkwe District in Otjozondjupa Region). As Dieckmann (2003: 59–60, 2007: 186, 189–191) notes, in the final report of the Bushman Commission, the Hai//om, the largest San population in the country, were not to receive a reserve. There were likely a number of reasons for this decision, some of them relating to the labor needs of commercial farmers and to the fears of some people in Nature Conservation that Hai//om might have a significant impact on the wildlife populations in the reserve.

In 1954, all but 12 Hai//om families who worked for Nature Conservation were told that they would have to leave the reserve. The other 400–500 Hai//om either had to resettle in Ovamboland or on white commercial farms south of the reserve (Widlok, 1999: 25–27; Gordon & Douglas, 2000: 165; Dieckmann, 2003: 60, 2007: 186ff.). According to Dieckmann (2007: 192), the Native Commissioner of Ovamboland told the Hai//om that they "had to leave the reserve for the sake of the game," and would be allowed to return only if they were in possession of a permit.

In the mid-1950s, the San and other peoples in Namibia were under the administrative oversight of the Department of Bantu Administration and Development. In 1954, the issuing of the South West African Native Affairs Administration Act laid out the bureaucratic structure under which San and other "native peoples" fell. In this system, the San had no right to self-representation, they had no leaders recognized formally by the South West African Administration, and they had no say about what could be done with regard to the land. The Hai//om did have leaders who were recognized by local Hai//om communities, but by and large the South West African administration did not utilize these leaders as intermediaries or as channels for communication to the Hai//om people. According to reports obtained during the course of oral historical investigations, Hai//om leaders were, in fact, consulted on occasion by colonial authorities, and there are indications in letters and reports in the Namibia National archives that local government officials, police, and others had contacts with Hai//om leaders and community members. There are some Hai//om who live in settlements created during the apartheid era (for example, during the liberation struggle in the 1970s and 1980s) and around mission stations (Widlok, 1999: 4–5).

Table 3.4 National Parks and Game Reserves in Namibia

National Park or Game Reserve	Date of Founding	Size (in sq km)
Ai-Ais/Richtersveld Transfrontier Park (including Fish River Canyon)	August 1, 2003	5086 km²
Brandberg National Monument (World Heritage Site)	October 3, 2002	650 km²
Bwabata National Park (formerly, West Caprivi Game Reserve)	2003 (1968-West Caprivi)	5715 km²
Dorob National Park	December 1, 2010	57,722 km²
Etosha National Park (formerly, Etosha Game Reserve)	1907 (93,240 km²) reduced in size in 1975	22,912 km²
Kaudum National Park (formerly Kaudum Game Reserve)	1989 declared game reserve, 2007 declared a national park	3841 km²
Madumu National Park	1990 (Zambezi Region)	1010 km²
Mahango Game Reserve	1989	244.6 km²
Mamili National Park	1990	320 km²
Namib-Naukluft National Park	1907 (1966-Naukluft Mountain Zebra Park)	49,768 km²
Namib Sand Sea	2013 (World Heritage Site)	30,777 km² Buffer zone 8995 km²
Nkasa Rupara National Park	Mini-Okavango, 430 bird spp., canoe trails	320 km²
Skeleton Coast National Park	1972	17,400 km²
Sperrgebiet National Park	2004	26,000 km²
Twyfelfontein (/Ui-//aes)	1952 (WHS in 2007)	42.69 km²
Waterberg Game Reserve	1972	405.5 km²
Totals	16 protected areas	261,760.8 km²
Country—824,292 km²	14% of the country	31% of the country

Note: Data obtained from the Ministry of Environment, Forestry, and Tourism, Namibia, www.met. gov.na/maps/attractions.htm

The political and land situations of the Hai//om have long been complex. When the Odendaal Commission recommended the creation of "Bushmanland" (now part of Otjozondjupa region) along with other ethnic "homelands" (e.g. Hereroland, Damaraland), in the 1960s, the Hai//om were omitted. As a consequence, the Hai//om were left largely landless. It is this history of dispossession and marginalization that has led to the contemporary Namibian government decisions to provide the Hai//om with land and development assistance (for additional information, see Suzman, 2001a, b; Harring & Odendaal, 2006a, b, 2007; Hitchcock, 2016; Koot & Hitchcock, 2019).

Etosha National Park is part of the traditional ancestral territory of the Hai//om (Dieckmann, 2003: 76). Etosha is one of approximately 16 protected areas in Namibia (see Table 3.4). At 22,912 km², it is one of the larger parks in the country and the one that hosts the largest number of international visitors (some 220,000 people per year). Hai//om have lived Etosha from 'time immemorial" as they put it, and they were there at the time of Etosha's establishment as a game park in 1907 (Dieckmann, 2001, 2003, 2007, 2009). Hai//om continued to live and work in Etosha until 1953–54, when the decisions were made by the South West African adminis-tration to remove them. Oral history evidence suggests that Hai//om who had been relocated out of the protected area continued to visit the park quietly from the 1950s through the present, sometimes to see relatives and to collect wild resources or visit the graves of loved ones.

The removals were considered necessary because it was argued the Hai//om allegedly engaged in such acts as begging from tourists and disturbing wildlife and tourists who were visiting at water-holes (Friederich, 2014). In the 1950s, European farmers, aided by some Hai//om who knew Etosha intimately, engaged in hunting in the south eastern section of the park (Berry, 1997: 4). Etosha's Chief Nature Conservator, Peter Stark, countered this effectively by rounding up suspected poachers, prosecuting the leaders, and taking the Hai//om into service in Etosha as trackers (Friederich, 2014). Over the the next 60 years, Hai//om were prevented from entering Etosha or were discouraged from going into the park unless they were seeking employment; a small number of San were able to obtain jobs at the tourist facilities in Etosha (Okaukuejo, Halali Rest Camp, and Namutoni).

Beginning in 1928, Hai//om and other people were prevented from carrying traditional weapons (bows, arrows, spears, knives) which led to arrests for contra-vening the law banning firearms and other weapons (Friederich, 2014: 60). Some Hai//om accumulated livestock, which the government authorities felt were inap-propriate in a protected area. Government authorities who came into power in 1948 began imposing increasingly restrictive measures on livestock ownership by Hai//om (Friederich, 2014: 60–61). The formal eviction of the Hai//om and their livestock from Etosha occurred in 1954–1955 (Dieckmann, 2007; Friederich, 2014: 60–67), based on an eviction notice issued by the South West African administrtion (see Friederich, 2014: 68–69).

In 2007, the Namibian government again sought to relocate additional Hai//om outside of Etosha in some farms south of the national park. As of 2023, some 600 people had been resettled on the farms that had been set aside for them just south of Etosha National Park, A section of Etosha was allocated to the Hai//om as part of the Gobaub Concession (Roger Collinson, Gerson Kamatua, personal com-munications, 2022). The government of Namibia has been monitoring the progress of the Hai//om resettlement as part of its efforts to promote development for 'marginalized communities' (Lawry et al., 2012; Dieckmann, 2014, 2020; Division of Marginalized Communities, 2018; Koot & Hitchcock, 2019; Odendaal et al., 2020).

In September 2015, the Hai//om filed a collective action lawsuit against the government of Namibia, seeking to clarify the rights of the Hai//om in and around Etosha National Park (High Court of Namibia, 2015). This case was rejected by the High Court on 28 August, 2019 (High Court of Namibia, 2019). Subsequently, an appeal was filed with the High Court of Namibia. This appeal was rejected by the Appeals Court of Namibia in March 2022 (Menges, 2022) Meetings were being held with the applicants in the collective action lawsuit along with their communities in the latter part of 2022 and through the first few months of 2023 (Willem Odendaal, personal communication, 4 April 2023). The Hai//om are concerned that their rights are not being recognized, and that the High Court rejected their legal case on false grounds. Hai//om activism continues to be important in Namibia and internationally.

Case 4: Bwabwata National Park, Namibia

Bwabwata National Park, formerly the West Caprivi Game Reserve, was established in what was then South West Africa (now Namibia) in 1963. Some Khwe San in the area were resettled outside of the reserve. Some of the Khwe moved to what is now the Kavango Region, and others moved into Botswana. In 1968, the South West African administration erected a fence along the Botswana-Namibia border in West Caprivi. Some Khwe were relocated away from the border, and a number of Khwe from South West Africa moved into northern Botswana. Fences in Namibia have had a significant effect on wildlife population movements, which has sometimes led to a decline in wildlife numbers, especially antelopes, but also giraffes and, in some cases, megafauna like elephants.

In 1970 !Xun and Khwe in southern Angola were caught in the fighting between the MPLA (one of the major liberation movements in Angola) and the Portuguese. Some of the Khwe who survived fled into Zambia and sought refugee status. In 1972 South African Police (SAP) units were established in the Caprivi Strip region as part of the efforts to contain the South West African Peoples Organization during the liberation struggle that lasted from 1965 to 1990. In 1974 Bushman Battalion 31 was established by South African Defense Force (SADF) in the Caprivi Strip, replacing the South African Police. As it worked out, Portuguese forces withdrew from Angola after a coup in Portugal in 1974. South African Defense Force (SADF) military camps were established at Alpha (later called Omega) and Buffalo in West Caprivi.

In 1976 additional South African Defense Force military camps were established in the Caprivi Strip (e.g. Chetto, Dodge City). The majority of the Khwe and !Xun San were concentrated in a few places in West Caprivi during this period. In 1989, after years of fighting, South African Defense Force troops were withdrawn from West Caprivi. Namibian Independence was celebrated on March 21st, 1990. Some 4500 !Xun and Khwe San, former SADF soldiers and their families, were moved from Namibia to South Africa and resettled at Schmidtsdrift, west of Kimberley. The !Xun and Khwe San now have their own land at Platfonteein in South Africa (Hitchcock, 2012; Taylor, 2012). The !Xun and Khwe in Namibia, on the other hand, continue to live in limbo as far as land tenure rights and political representation is concerned (Dieckmann, 2014; Paksi & Pyhälä, 2018; Boden, 2020).

In April, 1997, the government of Namibia's Ministry of Prisons announced plans to expand their operations and establish a prison farm on the Okavango river near Popa Falls, a prime tourism site that was already being utilized by the Khwe for a community campsite known as N//goavaca. This action was protested by the

Working Group of Indigenous Minorities in Southern Africa (WIMSA), Integrated Rural Development and Nature Conservation, and the National Society for Human Rights (NSHR) of Namibia. After extensive efforts to negotiate a settlement, the Legal Assistance Centre (LAC) of Namibia filed a legal case against the Ministry of Prisons in the High Court of Namibia. The LAC was represented by lawyer Peter Watson. An anthropologist, Ina Orth, who had worked in West Caprivi in January–February, 1998, provided an affidavit in the case. Eventually, after intensive discussions, the Ministry of Prisons decided to move the prison farm, and the legal case was settled out of court in August of 1998.

The West Caprivi Game Reserve was declared to be a national park (Bwabwata National Park) in 2003. The Khwe living in the park were allowed to remain inside of the park, although some of their activities were restricted such as hunting, grazing, fishing and collecting of wild plants (Taylor, 2012). Bwabwata National Park covers an area of 5715 km^2. It is now part of the Kavango Zambezi Transfrontier Conservation Area (KAZA) which is some 519,912 km^2 (200,739 mi^2) in size and incorporates five countries (Angola, Botswana, Namibia, Zambia, and Zimbabwe) (Cumming, 2008; Taylor, 2012).

The history of the Bwabwata and surrounding areas has not been without conflict and disagreement (Lenggenhager, 2018; Boden, 2020; Paksi, 2020). There were refugee flows of Khwe into Botswana of Khwe after what was described as a 'secessionist incident' in August, 1999 in what was then East Caprivi. In November, 1999, there were some 2300 Namibian refugees in Dukwe Refugee Camp in north eastern Botswana. Some of the refugees were from Angola and other places who came to Namibia were relocated to the Osire Refugee Camp in the central part of the country. This refugee camp's population grew substantially in the 1990s, but after 2002, with Peace Accords signed between Namibia and Angola, some of the refugees returned to Angola (Hitchcock, 2012). Today, in 2023, the Osire Refugee Camp has about 2000 refugees living in it.

No Khwe Traditional Authority (TA) has been agreed upon by the Government of Namibia. In addition to that, the land status of the western part of Zambezi Region and East Kavango Region is still under discussion (Paksi & Pyhälä, 2018; Paksi, 2020). What is unique about Bwabwata National Park is the fact that some Khwe and Mbukushu were allowed to reside there even after the declaration of the West Caprivi Game Reserve in 1963 and Bwabwata National Park in 2003. Neither group has the right to hunt in the park today. Safari companies are, however, allowed to bring clients into Bwabwata National Park. The residents of Bwabwata are allowed to keep livestock and plant gardens in the park. Bwabwata National Park in specifically designated zones, according to the Ministry of Environment, Forestry, and Tourism (MEFT).

The MEFT employs an integrated management approach in which the residents of the Park are supposed to be actively involved in park management.. The Ministry works closely with the official residents' organisation for the Park, the Kyaramacan Association (KA) (Paksi, 2020). The Kyaramacan Association in 2018 employed 40 Khwe ecological monitors, including 25 Community Game Guards (CGGs) and 15 Community Resource Monitors (CRMs), who carried foot-patrol across the

Table 3.5 Land Tenure Zoning in Namibia

Land tenure category	Size in square kilometers	Percentage of the country
Communal land (non-titled land)	287,204.43 km²	34.84
Commercial land (freehold, titled land)	397,283.64 km²	48.2
State land (parks, national monuments and game reserves, etc.)	139,064.37 km²	16.87
Resettlement land (portion of freehold land)	30,219 km²	0.76 of commercial land
Total	824,292 km²	100

Note: Data obtained from Mendelsohn et al. (2009), the Namibia Statistics Agency, the Ministry of Land Reform and Land Resettlement (MLRR), the Ministry of Environment, Forestry, and Tourism (MEFT), Government of the Republic of Namibia

Bwabwata National Park on a daily basis (Paksi, 2020). The numbers of employees of the Kyaramacan Association dropped significantly, however, after the declaation of the coronavirus pandemic lockdowns in Namibia in March 2020, and the numbers of safari hunting clients and ecotourists visiting the Park dropped by some 90–95% according to the Ministry of Environment, Forestry, and Tourism in Namibia. The numbers of tourists visiting northern Namibia began to pick up in 2021 and 2022. One of the issues of concern to the Khwe in Bwabwata is that most of the tourists who visit the park come to see wildlife, and cultural tourism remains poorly developed. The Khwe continue to raise their voices regarding their rights in the Bwabwata national Park (Van Wyk, 2022). Not only do they wish to have their security of tenure recognized, but they also want to have the Traditional Authority of the Khwe people recognized by the government of Namibia.

The Republic of Namibia has set aside almost 17% of its total land as national perks, game reserves, and national monuments (see Table 3.5). It has also created some 87 communal conservancies on communal land in the country. In these areas, communities have the right to oversee the ways in which wildlife is utilized and to work with the Ministry of Environment, Forestry, and Tourism to set quotas that are sustainable (Republic of Namibia, 1996). In many ways, Namibia has become a model for how to handle wildlife conservation and management.

Case 5: Kgalagadi Transfrontier Park, South Africa and Botswana

In the early 1930s, the residents of Kalahari Gemsbok Park in South Africa were told that they had to leave the park once it was established. A group of 84 families of Coloureds along with 5000 head of livestock were resettled south of the park in the northern Cape region (Van Wyk & Le Riche, 1984: 23–24). Initially, the San were not required to leave the park. As was noted by Van Wyk and Le Riche (1984),

In 1936 during a patrol through the central block of the park Joeple (Le Riche) encountered a group of Bushmen at Sewe Pan, as it was called then. Since it was Minister Grobler's express wish that the park should act as a refuge for the Bushmen to save them from extinction, this group of 20 were lured closer by means of tobacco and food to put them at ease. Although communication was impossible, they were persuaded by means of gestures to accompany the ranger to Gemsbokfontein where their needs could be provided for (Van Wyk & Le Riche, 1984: 25).

This is a classic example of voluntary isolated indigenous people (VIIPs) who were contacted and convinced to leave the area where they lived and move to a place where they could be resettled. In Botswana, !Xõó San and Nama were required to leave the Kalahahari Gemsbok Park in the 1930s. Several hundred ≠Khomani on the South African were told to leave the park in 1930–31 but they were not pressured to do so until the latter part of the 1930s (Holden, 2007; Puckett, 2018). A number of ≠Khomani ended up living on the peripheries of Kalahari Gemsbok Park, and some moved south to tourist camps in the Cedarberg Mountains, where they lived for several decades working at tourist camps. They returned to the Mier area and Upington in the early part of the 2000s.

The ≠Khomani San of South Africa filed a land claim under the new constitution in South Africa in 1994, seeking rights to benefits from the Kalahari Gemsbok Park as well as land rights and funds for development (Ellis, 2012). In 1999, the first phase of the claim was resolved out of court. The ≠Khomani were granted some 38,000 hectares of land in their ancestral area. They also got rights to some of the gate receipts from the visitors to the park. Subsequently, in another out-of court settlement, the ≠Khomani got co-management rights over land within the Kgalagadi Transfontier Park from which they had been evicted in 1931 (Chennells, 2002a, b, 2003; Chennells & du Toit, 2004). This settlement was approved by the South African Parliament on 31 August 2002. This was the first successful land claim of South Africa's San people (Holden, 2007). Since that time the community has struggled somewhat, with intragroup conflict and difficulties relating to leadership and distribution of economic benefits (Julie Grant, personal communication, December 2022). Nevertheless, the case remains an important one in South Africa, particularly in terms of recognizing communal land rights of San people.

An examination of the Kgalagadi Transfrontier Park section inside Botswana shows that San, Nama, and others who were living in the park in the 1930s were told to leave the park. Some of them moved north to the Kgalagadi Ghanzi border, where they resided for the next 90 years, until they were told to move again because of the establishment of the Western Kalahari Conservation Corridor, aimed at providing wildlife migration routes between the Kgalagadi Transfronteir Park and the Central Kalahari Game Reserve (see Fig. 3.1). There are currently no residents living inside of Kgalagadi Transfrontier Park, but some people remain on the peripheries and are able to get some of the benefits from the gate receipts from park visitors.

The western Kalahari is currently undergoing additional transformations as a result of a government of Botswana United Nations Development Programme, Global Environmental Facility financed ecological and community livelihoods project known as the KGDEP (see Government of Botswana and United Nations

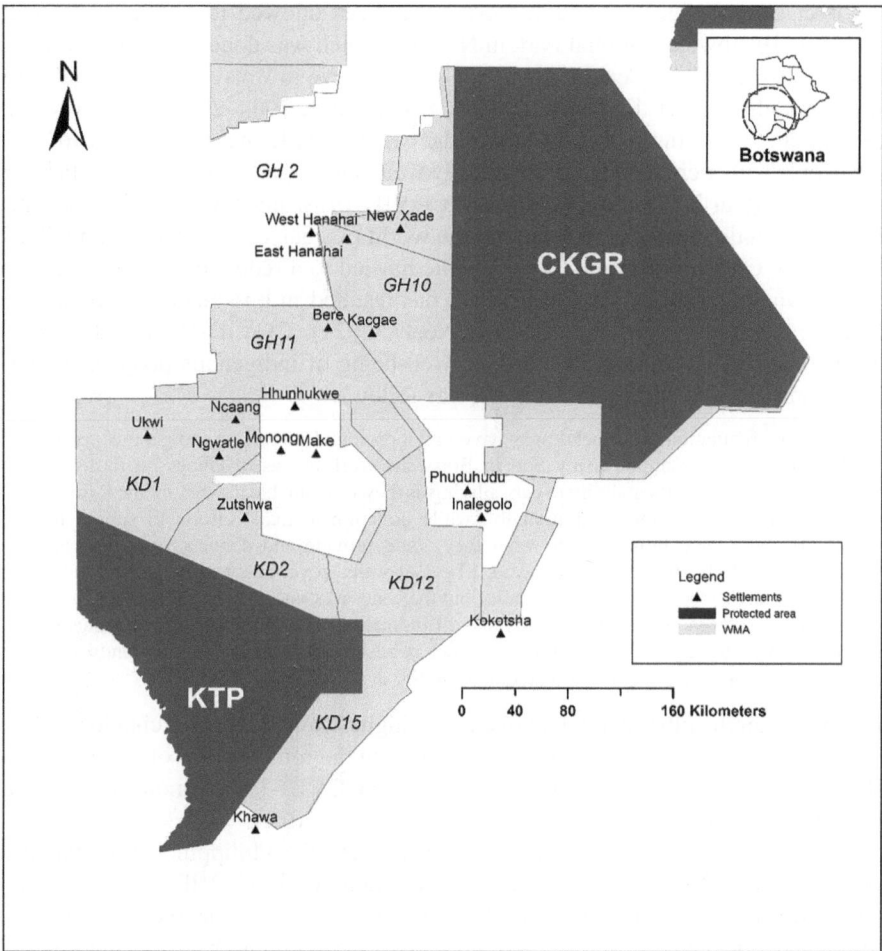

Fig. 3.1 Western Kalahari Protected Areas, Wildlife Management Areas, and Remote Area communities

Development Program, 2017). So far, no additional communities have been required to move under this latest development project, though some have relocated voluntarily in order to move away from high livestock areas.

Conclusions

Several conclusions can be drawn from these case studies. First, indigenous peoples have tended to be excluded from national parks and other protected areas when they were created. Park planning has resulted in the removals of some of their residents, but over time, indigenous people have either been allowed to return to the parks including Bwabwata National Park in Namibia which was done through negotiation (Paksi, 2020), and the Central Kalahari Game Reserve in Botswana after legal cases were brought against the government of Botswana in 2004–2006 and 2010–2011 (Sapignoli, 2018). In nearly all cases in the US and southern Africa, the majority of the people who were relocated involuntarily were hunter-gatherers and small farmers.

Beginning in 2020, the coronavirus (COVID-19) pandemic has had substantial impacts on indigenous peoples around the world (Kaplan et al., 2020; Mamo, 2022; Ro, 2023). On the one hand, the pandemic has led to a reduction in the number of tourists visiting indigenous areas, which has resulted in a reduction in impacts on biodiversity (Pearson et al., 2020; McNeely, 2021). COVID-19 has also led to changes in land use and the health and well-being of indigenous peoples. To take one example, Botswana, it was reported as follows:

> Pandemic regulations and lockdowns have had a profound impact on indigenous peoples' livelihoods. For example, many San in Botswana work as casual labour on farms. The lockdown impacted their ability to earn a living as they were unable to work on the farms due to movement restrictions. Additionally, while government made efforts to supply food rations, there were limitations to what they could provide and delivery to remote areas were particularly difficult and delayed. The Botswana government also made efforts to deliver water to the remote areas in tanks, but the water was not adequate. Given that access to water is always a challenge for almost all indigenous communities, the inadequacy of deliveries during COVID-19, coupled with lack of basic needs such as soap for handwashing was, and continues to be, life threatening (ARISA & USAID, 2020: 12).

Botswana, on the other hand, had one of the highest COVID-19 vaccination rates in the world. Vaccine resistance was not reported in the remote areas of the country. It should also be noted that the mortality rate from COVID-19 in remote communities was relatively low in Botswana (Hitchcock & Frost, 2022).

Governments ranging from Brazil to Nepal and the Philippines have failed to provide adequate assistance to indigenous communities for COVID-19 relief. Part of the reason for this situation is the differentiation of remote people from those in towns and cities, who have received more assistance ranging from vaccinations and masks to provision of soap for handwashing and food. In the Amazon Basin, indigenous communities were hard-hit by COVID-19 (Kaplan et al., 2020). In some cases, food and water were provided to indigenous communities but the vaccination rates have been relatively low.

The most common arguments made by governments and wildlife biologists have been that indigenous peoples tend to overexploit the wildlife and other natural resources in the areas where they reside (see, for example, Owens & Owens, 1981; Spinage, 1991). There are also arguments that indigenous peoples and their domestic animals are a threat to wildlife, as argued, for example, in the first Central Kalahari Game Reserve legal case (Sapignoli, 2018: 226, 233). While indigenous peoples are sometimes blamed for poaching and for over-exploitation of resources, there is mounting evidence that this is not generally the case (see, for example, Hitchcock et al., 2020).

Indigenous peoples employ numerous traditional and contemporary natural resource management strategies, including diversification of their exploitation of resources as numbers decline, purposeful management of resources using fire and transplantation of desirable plants, and engineering of the natural landscape in various ways in order to prevent over-exploitation of resources. Indigenous peoplesuing for the demilitarization of protected areas, and they have sought to ensure that there is no impunity for people who torture and kill residents of protected areas in the name of anti-poaching. One suggestion that they have made is that indigenous peoples themselves be allowed to oversee protected areas and that they be allowed to employ non-coercive tactics to ensure that protected areas are free of human rights abuses. Indigenous peoples in protected areas also want to receive some of the benefits of tourism and to be able to share their knowledge, stories, and knowledge with visitors, scientists, and park authorities.

References

ARISA and USAID. (2020). The impact of COVID-19 on the rights of indigenous peoples in Southern Africa. Washington, D.C. U.S. Agency for International Development and Johannesburg, South Africa: Advancing Rights in Southern Africa.

Bechuanaland Protectorate. (1961). *The fauna conservation proclamation*. Bechuanaland Protectorate Administration.

Bechuanaland Protectorate. (1963). *The central Kalahari game reserve regulations*. Bechuanaland Protectorate Administration.

Berry, H. H. (1997). Historical review of the Etosha region and its subsequent administration as a National Park. *Madoqua, 20*(1), 3–12.

Boden, G. (2020). Land and resource rights of the Khwe in Bwabwata National Park. In W. Odendaal & W. Werner (Eds.), *"Neither here nor there": Indigeneity, marginalisation, and land rights in post-Independence Namibia* (pp. 229–254). Legal Assistance Centre.

Botswana Court of Appeal. (2011). In the court of appeal of Botswana held at Lobatse. *Court of Appeal No. CACLB-074-10. High Court Civil Case No. MAHLB 000 393-09 In the matter between Matsipane Mosetlhanyene, First Appellant, and Gakenyatsiwe Matsipane, Second Appellant, and the Attorney General Respondent. Heard 17 January, 2011 and delivered 27 January, 2011.* Court of Appeal.

Boustany, N. (1995, December 18). The Bushmen's Advocate: Straddling two worlds, John Hardbattle speaks to the plight of the No'akwe of Botswana. *Washington Post*, D1, D4.

Carruthers, J. (1995). *The Kruger National Park: A social and political history*. University of KwaZulu Natal Press.

Catholic Commission for Justice and Peace in Zimbabwe and Legal Resources Foundation. (2008). *Gukurahundi in Zimbabwe: A Report on the Disturbances in Matabeleland and the Midlands 1980–1988*. Columbia University Press.

Chennells, R. (2002a). The /Khomani San land claim. *Cultural Survival Quarterly, 26*(1), 51–52.

Chennells, R. (2002b). *The ‡Khomani san land claim*. Paper presented at the indigenous rights in the commonwealth project, Africa regional expert meeting; indigenous peoples of Africa coordinating committee (IPACC). Cape Town.

Chennells, R. (2003). Case study 9: South Africa: The ‡Khomani San of South Africa. In J. Nelson & L. Hossack (Eds.), *Indigenous peoples and protected areas in Africa: From principle to practice* (pp. 269–293). Forest Peoples Programme.

Chennells, R., & du Toit, A. (2004). The rights of indigenous peoples in South Africa. In R. K. Hitchcock & D. Vinding (Eds.), *Indigenous peoples' rights in southern Africa* (pp. 98–113). International Work Group for Indigenous Affairs.

Child, G. (1995). *Wildlife and people: The Zimbabwean success*. The Wisdom Foundation.

Cook, A., & Sarkin, J. (2009). Who is indigenous? Indigenous rights globally, in Africa, and among the San in Botswana. *Tulane Journal of International and Comparative Law, 18*, 93–130.

Cumming, D. H. M. (2008). *Planning and priorities for the Kavango-Zambezi transfrontier conservation area*. Conservation International.

Davison, T. (1977). *Wankie: The story of a great game reserve*. Irwin Press.

Dieckmann, U. (2001). 'The vast white place': A history of the Etosha National Park in Namibia and the Hai//om. *Nomadic Peoples, 5*(2), 125–153.

Dieckmann, U. (2003). The impact of nature conservation on San: A case study of Etosha National Park. In T. Hohmann (Ed.), *San and the state: Contesting land, development, identity, and representation* (pp. 37–86). Rudiger Koppe Verlag.

Dieckmann, U. (2007). *Hai//om in the Etosha region: A history of colonial settlement, ethnicity, and nature conservation*. Basler Afrika Bibliographien.

Dieckmann, U. (2009). *Born in Etosha: Homage to the cultural heritage of the Hai//om*. Legal Assistance Centre.

Dieckmann, U. (2014). Kunene, Oshana and Oshikoto regions. In U. Dieckmann, M. Thiem, E. Dirkx, & J. Hays (Eds.), *"Scraping the pot": San in Namibia two decades after independence* (pp. 173–232). Legal Assistance Centre and Desert Research Foundation of Namibia.

Dieckmann, U. (2020). From colonial land dispossession to the Etosha and Mangetti west land claim – Hai//om struggles in independent Namibia. In W. Odendaal & W. Werner (Eds.), *"Neither here nor there": Indigeneity, marginalisation, and land rights in post-Independence Namibia* (pp. 95–120). Legal Assistance Center.

Division of Marginalized Communities. (2018). Draft white paper on the rights of indigenous peoples in Namibia. :. Division of Marginalized Communities, Office of the President; Division of Marginalized Communities.

Dzangerai, M. (1995). The Batwa of western Rhodesia. *Nada, 54*, 6–14.

Ellis, W. N. (2012). *Genealogies and narratives of San authenticities. The ≠Khomani San land claim in the southern Kalahari*. PhD. Dissertation, University of the Western Cape.

Friederich, R. (2014). Etosha: Hai//om Heartland: Ancient Hunter-Gatherers and their environment. In H. Lempp (Ed.). Namibia Publishing House.

Gordon, R. J., & Douglas, S. S. (2000). *The Bushman Myth: The making of a Namibian underclass* (2nd ed.). Westview Press.

Government of Botswana. (1985). *Report of the Central Kgalagadi reserve fact finding mission*. Government Printer.

Government of Botswana and United Nations Development Programme. (2017). *Botswana project document: Managing the human-wildlife interface to establish the flow of agro-ecosystem services and prevent illegal wildlife trafficking in the Kgalagadi and Ghanzi Drylands*. Government of Botswana and United Nations Development Programme.

Güldemann, T. (2008). A linguist's view: Khoe-Kwadi speakers as the earliest food-producers of southern Africa. *Southern African Humanities, 20*, 93–132.

Güldemann, T. (2014). 'Khoisan' linguistic classification today. In T. Guldemann & A.-M. Fehn (Eds.), *Beyond 'Khoisan': Historical relations in the Kalahari Basin* (pp. 1–44). John Benjamins Publishing.

Hahn, C. H. L., Vedder, H., & Fourie, L. (Eds.). (1928). *The native tribes of South West Africa*. South West African Administration.

Harring, S., & Odendaal, W. (2006a). *'Our Land They Took': San Land Rights under Threat in Namibia*. Land, Environment and Development [LEAD] Project, Legal Assistance Center.

Harring, S., & Odendaal, W. (2006b). *'One Day We Shall All Be Equal': A socio-legal perspective on the Namibian land reform and resettlement*. Legal Assistance Center.

Harring, S., & Odendaal, W. (2007). *"No Resettlement Available": An assessment of the expropriation principle and its impact on land reform in Namibia*. Legal Assistance Center.

Haynes, G. (n.d.). *Hwange National Park: The Forest with a desert heart*. Manuscript in author's possession, University of Nevada-Reno.

High Court of Botswana. (2006). *Case No. MISCA 52/2002 in the Matter Between Roy Sesana, First Applicant, Keiwa Setlhobogwa and 241 others, Second and Further Applicants, and the Attorney General (in his capacity as the recognized agent of the Government of the Republic of Botswana). Judgment coram Hon. Mr. Justice M. Dibotelo, Hon. Justice U. Dow, Hon. Mr. Justice M. P. Phumaphi. 13 December, 2006*. High Court of Botswana.

High Court of Namibia. (2015). Case No A 201/15. In the High Court of Namibia, Main division, Windhoek, in the matter between Jan Tsamib and 7 others and the government of the Republic of Namibia and 13 others on the matter of the Hai//om People and their land. High Court of Namibia.

High Court of Botswana. (2002). *Central Kalahari Legal Case No. MISCA 52/2002 in the Matter Between Roy Sesana, First Applicant, Keiwa Setlhobogwa and 241 others, Second and Further Applicants, and the Attorney General (in his capacity as the recognized agent of the Government of the Republic of Botswana)*. High Court of Botswana.

High Court of Namibia. (2019). *Tsumib v government of the Republic of Namibia* (A 206/2015) [2019] NAHCMD 312 (21 August 2019). High Court of Namibia.

Hitchcock, R. K. (1982). *The Ethnoarchaeology of Sedentism: Mobility strategies and site structure among foraging and food producing populations in the eastern Kalahari Desert, Botswana*. PhD dissertation, University of New Mexico, Albuquerque.

Hitchcock, R. K. (1988). *Monitoring, research, and development in the remote areas of Botswana*. Remote Area Development Program and Ministry of Local Government and Lands and Norwegian Agency for International Development.

Hitchcock, R. K. (1995). Centralization, resource depletion, and coercive conservation among the Tyua of the northeastern Kalahari. *Human Ecology, 23*(2), 169–198.

Hitchcock, R. K. (2000). Traditional African wildlife utilization: Subsistence hunting, poaching, and sustainable use. In H. H. T. Prins & J. G. Grootenhuis (Eds.), *Conservation of wildlife by sustainable use* (pp. 389–415). Chapman and Hall.

Hitchcock, R. K. (2001). 'Hunting is our heritage': The struggle for hunting and gathering rights among the San of Southern Africa. In D. G. Anderson & K. Ikeya (Eds.), *Parks, property, and power: Managing hunting practice and identity within state policy regimes* (pp. 139–156). Senri Ethnological Studies 59. National Museum of Ethnology.

Hitchcock, R. K. (2002). 'We are the first people': Land, natural resources, and identity in the Central Kalahari, Botswana. *Journal of Southern African Studies, 28*(4), 797–824.

Hitchcock, R. K. (2012). Refugees, resettlement, and land and resource conflicts: The politics of identity among! Xun and Khwe San of Northeastern Namibia. *African Study Monographs, 33*(2), 73–132.

Hitchcock, R. K. (2016). Hunter-Gatherers, Herders, Agropastoralists, and Farm Workers: Hai//om and Ju/'hoansi San and their neighbors in Namibia in the 20th and 21st centuries. In K. Ikeya & R. K. Hitchcock (Eds.), *Hunter-gatherers and their neighbours in Asia, Africa, and South America* (pp. 263–284). Senri Ethnological Studies No. 94. National Museum of Ethnology.

Hitchcock, R. K., & Frost, J. (2022). Botswana. In D. Mamo (Ed.), *The indigenous world 2022* (pp. 39–48). International Work Group for Indigenous Affairs.

Hitchcock, R. K., Winer, N., & Kelly, M. C. (2020). Hunter-gatherers, farmers, and environmental degradation in Botswana. *Conservation and Society, 18*(3), 226–237.

Hitchcock, R. K., Begbie-Clench, B., & Murwira, A. (2016). *The San in Zimbabwe: Livelihoods, land, and human rights*. International Work Group for Indigenous Affairs (IWGIA), Open Society Initiative for Southern Africa (OSISA), University of Zimbabwe.

Hitchcock, R. K., & Nangati, F. M. (1992). *Assessment of the community-based resource utilization component of the Zimbabwe natural resources management project (690–0251)*. U.S. Agency for International Development.

Hitchcock, R. K., & Nangati, F. M. (1993). Drought, environmental change, and development among the Tyua of Western Zimbabwe. *International Work Group for Indigenous Affairs Newsletter, 2*(93), 42–46.

Hitchcock, R. K., & Nangati, F. M. (2000). People of the two-way river: Socioeconomic change and natural resource management in the Nata River Region of Southern Africa. *Botswana Notes and Records, 32*, 1–21.

Hitchcock, R. K., & Masilo, R. R. B. (1995). *Subsistence hunting and resource rights in Botswana.* Department of Wildlife and National Parks.

Holden, P. (2007). Conservation and human rights—The case of the Khomani San (Bushmen) and the Kgalagadi Transfrontier Park, South Africa. *Policy Matters, 15*, 57–68.

Ikeya, K. (2001). Some changes among the San under the influence of relocation plan in Botswana. In D. G. Anderson & K. Ikeya (Eds.), *Parks, property, and power: Managing hunting practice and identity within state policy regimes* (pp. 183–198). Senri Ethnological Studies No. 59. National Museum of Ethnology.

INK Center for Investigative Journalism. (2016). *The "shoot to kill policy" in pursuit of the truth: The story of Botswana's brutality.* INK Center for Investigative Journalism. Accessed December 10, 2020, from www.inkjournalism.org

Kaplan, H. S., Trumble, B. C., Stieglitz, J., Mamany, R. M., Cayuba, M. G., Moye, L. M., Alami, S., Kraft, T., Gutierrez, R. Q., Adrian, J. C., Thompson, R. C., Thomas, G. S., Michalik, D. E., Rodriguez, D. E., & Gurven, M. D. (2020). Voluntary collective isolation as a best response to COVID-19 for indigenous populations? A case study and protocol from the Bolivian Amazon. *The Lancet, 395*, 1727–1734.

Kiema, K. (2010). *Tears for my land: A social history of the Kua of the central Kalahari game reserve, Tc'amnqoo.* Bay Publishing.

Koot, S., & Hitchcock, R. K. (2019). In the way: Perpetuating land dispossession of the indigenous Hai//om and the collective action lawsuit for Etosha National Park and Mangetti West, Namibia. *Nomadic Peoples, 3*(1), 55–77.

Laverick, J. (2015). *The Kalahari Killings: The true story of a wartime double murder in Botswana, 1943.* The History Press.

Lawry, S., Begbie-Clench, B., & Hitchcock, R. K. (2012). *Hai//om resettlement farms: Strategy and action plan.* Ministry of Environment and Tourism and Millennium Challenge Corporation.

Lenggenhager, L. (2018). *Ruling nature, controlling people: Nature conservation, development, and War in North Eastern Namibia since the 1920s.* Baslet Afrika Bibliographien.

Mamo, D. L. (2022). *The Indigenous World 2022.* International Work Group for Indigenous Affairs.

Mberengwa, I. (2000). *The communal areas management program for indigenous resources (CAMPFIRE) and rural development in Zimbabwe's marginal areas: A study in sustainability.* Ph.D. Dissertation, University of Nebraska.

McNeely, J. A. (2021). Nature and COVID 19: The pandemic, the environment, and the way ahead. *Ambio, 50*, 767–781.

Mendelsohn, J., Jarvis, A., Roberts, C., & Robertson, T. (2009). *Atlas of Namibia: A portrait of the land and its people* (3rd ed.). Sunbird Publishers.

Menges, W. (2022). San group appeal in Etosha rights case fails. *The Namibian*, 19 March 2022.

Mogomotsi, E. J., & Madigele, P. K. (2017). Live by the gun, die by the gun: Botswana's 'shoot to kill' policy as an anti-poaching strategy. *South Africa Crime Quarterly, 60*, 51–59.

Mogwe, A. (1992). *Who Was (T)here First? An assessment of the human rights situation of Basarwa in selected communities in the Gantsi District, Botswana.* Botswana Christian Council.

Morinville, C., & Rodina, L. (2013). Rethinking the human right to water: Water access and dispossession in Botswana's central Kalahari game reserve. *Geoforum, 49*, 150–159.

Muboko, N., Muphoshi, V., Tarakini, T., Gandiwa, E., Vengesayi, S., & Makuwe, E. (2014). Cyanide poisoning and African elephant mortality in Hwange National Park, Zimbabwe, a preliminary assessment. *Pachyderm, 55*, 92–94.

Muboko, N., Gandiwa, E., Muposhi, V., & Tarakini, T. (2016). Illegal hunting and protected areas: Tourist perceptions on wild animal poisoning in Hwange National Park, Zimbabwe. *Tourism Management, 52*, 170–172.

Mutowo, M. (2001). Animal diseases and human populations: The rinderpest epidemic in Zimbabwe 1896-1898. *Zambezia, 28*(1), 1–22.

Ndlovu, D., Begbie-Clench, B., Hitchcock, R. K., & Kelly, M. C. (2022). The Tshwa San of Zimbabwe: Land, livelihoods, and ethnicity. In K. Helliker, J. Matanzima, & P. Chadambuka (Eds.), *Livelihoods and ethnicity in Zimbabwe* (pp. 31–50). Springer.

Ng'ong'ola, C. (2007). Sneaking aboriginal title into Botswana's legal system through a side door: Review of *Sesana and others v. the Attorney General. Botswana Law Journal, 6*, 103–123.

Ngwenya, D. (Ed.). (2018). *Healing the wounds of Gukurahundi in Zimbabwe: A participatory action project.* Springer.

Odendaal, W., Gilbert, J., & Vermeylen, S. (2020). Recognition of ancestral land claims for indigenous peoples and marginalized communities in Namibia: A case study of the Hai//om legislation. In W. Odendaal & W. Werner (Eds.), *"Neither here nor there": Indigeneity, marginalisation, and land rights in post-Independence Namibia* (pp. 121–142). Legal Assistance Centre.

Owens, M., & Owens, D. (1981). *Preliminary final report on the central Kalahari predator research project.* Report to the Department of Wildlife and National Parks.

Paksi, A. (2020). *'Survivng Development': Rural development interventions, protected area management, and formal education with the Khwe San in Bwabwata National Park, Namibia.* Faculty of Social Sciences, University of Helsinki.

Paksi, A., & Pyhälä, A. (2018). Pursuing employment opportunities - The Khwe San in Bwabwata National Park, Namibia. In F. R. Puckett & K. Ikeya (Eds.), *Research and activism among the Kalahari San Today: Ideals, challenges, and debates* (pp. 199–218). Senri Ethnological Studies 99. National Museum of Ethnology.

Pearson, C. J., Macquarie, J. W., & Smith, A. C. (2020). *COVID-19 Impacts on indigenous people in Africa.* Unpublished Manuscript.

Pratchett, L. J. (2020). Language contact and change in eastern Botswana: New insights from the pronominal system of an undocumented Kalahari Khoe language. *Language in Africa, 1*(1), 34–64.

Pratchett, L. J. (2021). *Centering the periphery: A comparative analysis of pronominal systems and the implications for Kalahari-Khoe classification.* Unpublished Manuscript in possession of the author.

Puckett, R. F. (2018). 'The space to be themselves': Confronting the mismatch between South Africa's land reform Laws and Traditional San organization among the Khomani. In F. R. Puckett & K. Ikeya (Eds.), *Research and activism among the Kalahari San today: Ideals, challenges, and debates* (pp. 283–342). Senri Ethnological Studies 99. National Museum of Ethnology.

Rankomise, A. O. (2015). *The Tshwa: Zimbabwe's forgotten people.* Konrad-Adenauer-Stiftung.

Republic of Namibia. (1996). *Nature Conservation Amendment Act.* No. 5 of 1996. Windhoek. Government Printer.

Ro, C. (2023). Legacy of COVID-19 for indigenous health in the Brazilian Amazon. *British Medical Journal, 380*, o3005. https://doi.org/10.1136/bmj.o3005

Sapignoli, M. (2015). Dispossession in the age of humanity: Human rights, citizenship, and indigeneity in the central Kalahari. *Anthropological Forum: A Journal of Social Anthropology and Comparative Sociology, 25*(3), 285–305.

Sapignoli, M. (2016). Indigenous mobilization and activism: The San, the Botswana state, and the international community. In C. Lennox & D. Short (Eds.), *Handbook of indigenous peoples' rights* (pp. 268–281). Routledge.

Sapignoli, M. (2018). *Hunting justice: Displacement, law, and activism in the Kalahari.* Law and society series. Cambridge University Press.

Sapignoli, M., & Hitchcock, R. K. (2013). Indigenous peoples in southern Africa. *The Round Table: The Commonwealth Journal of international Affairs, 102*(4), 355–365.

Sapignoli, M. (2009). Indigeneity and the expert: Negotiating identity in the case of the Central Kalahari Game Reserve. In M. Freeman & D. Napier (Eds.), *Law and anthropology* (pp. 247–268). Oxford University Press.

Sarkin, J., & Cook, A. (2010–2011). The human rights of the San (Bushmen) of Botswana – The clash of the rights of indigenous communities and their access to water with the rights of the state to environmental conservation and mineral resource exploitation. *Journal of Transnational Law and Policy, 20*, 1–40.

Saugestad, S. (2001). *The inconvenient indigenous: Remote area development in Botswana, donor assistance, and the first people of the Kalahari.* Nordic Africa Institute.

Saugestad, S. (2011). Impact of international mechanisms on indigenous rights in Botswana. *The International Journal of Human Rights, 15*(1), 37–61.

Schoeman, P. J. (1951a). *Jagters van die Woestylnland (hunters of the desert land).* Timmins.

Schoeman, P. J. (1951b). *Voorlopije agters van die Kommisiee vir die nehoud van die Boesmanbevolking in Sudwes-Africa, 1950 (Interim Report).* South West Africa Administration.

Schoeman, P. J. (1953). *Report of the commission on the study of the Bushmen.* South West Africa Administration.

Scoones, I., Marongwe, N., Mavedzenge, B., Mahenehene, J., Murimbarimba, F., & Sukume, C. (2011). *Zimbabwe's land reform: Myths and realities.* James Currey, Weaver Press, Boydell & Brewer, Inc. and Jacanda Media.

Seleka, T. B., Siphambe, H., Ntseana, D., Mbere, N., Kerapeletswe, C., & Sharp, C. (2007). *Social safety nets in Botswana: Administration, targeting, and sustainability.* Lightbooks and Botswana Institute for Development Policy Analysis (BIDPA).

Silberbauer, G. B. (1965). *Report to the government of Bechuanaland on the Bushman survey.* Bechuanaland Government.

Silberbauer, G. B. (1981). *Hunter and habitat in the Central Kalahari Desert.* Cambridge University Press.

Silberbauer, G. B. (2012). Why the central Kalahari game reserve? *Botswana Notes and Records, 44*, 201–203.

Skidmore-Hess, C. (2021). Murder in Nata: Landscapes of colonial justice and authority in colonial Bechuanaland. *Journaol of Colonialism and Colonial History.*

Solway, J. (2009). Human rights and NGO 'wrongs': Conflict diamonds, culture wars, and the 'Bushman question'. *Africa, 79*(3), 329–343.

Spinage, C. (1991). *History and evolution of the Fauna conservation Laws of Botswana.* The Botswana Society.

Suzman, J. (2001a). *An introduction to the regional assessment of the status of the San in Southern Africa.* Legal Assistance Center.

Suzman, J. (2001b). *An assessment of the status of San in Namibia.* Legal Assistance Center.

Suzman, J. (2004). Etosha dreams: An historical account of the Hai//om predicament. *Journal of Modern African Studies, 42*(2), 221–238.

Tanaka, J. (1980). *The San, hunter-gatherers of the Kalahari. A study in ecological anthropology.* Tokyo University Press.

Tanaka, J. (1987). The recent changes in the life and Society of the Central Kalahari San. *African Study Monographs, 7*, 37–51.

Tanaka, J. (2014). *The bushmen: A half-century Chronicle of Transformation in hunter-gatherer life and ecology* (M. Sato, Trans.). Kyoto University Press and Trans Pacific Press.

Tanaka, J., & Sugawara, K. (2010). *An encyclopedia of /Gui and//Gana culture and society.* Laboratory of Cultural Anthropology, School of Human and Environmental Studies, Kyoto University.

Taylor, J. J. (2009). Differentiating "Bushmen" from "Bantus": Identity building in West Caprivi, Namibia, 1930-1989. *Journal of African History, 50*(3), 417–436.

Taylor, J. J. (2012). *Naming the land: San identity and community conservation in Namibia's West Caprivi*. Demasius Publications and Basler Afrika Bibliographien.

Van Onselen, C. (1972). Reactions to rinderpest in Southern Africa, 1896-97. *Journal of African History, 13*(3), 473–478.

Van Wyk, C. (2022). *Bwabwata National Park – The Khwe must be heard*. Windhoek: Legal Assistance Center, 5 April 2022.

Van Wyk, P., & LeRiche, E. A. N. (1984). The Kalahari Gemsbok Park: 1931-1981. *Koedoe, 1984*, 21–31.

Vogelsang, R. (2005). The past of the Etosha National Park: Oral history and archaeological evidence. *Nyame Akuma, 63*, 2–4.

Widlok, T. (1999). *Living on Mangetti: 'Bushman' Autonomy and Namibian Independence*. Oxford University Press.

Workman, J. L. (2009). *Heart of dryness: How the last bushmen can help us endure the coming age of permanent drought*. Walker and Company.

World Bank. (2015). *Botswana poverty assessment*. The World Bank.

Zips-Mairitsch, M. (2013). *Lost Land? (Land) Rights of the San in Botswana and the Legal Concept of Indigeneity in Africa*. Lit Verlag and Copenhagen: International Work Group for Indigenous Affairs.

Chapter 4
Social Impacts of Conservation-Forced Resettlement

Introduction

In this chapter we consider conservation-forced resettlement (CFR) and its social impacts.

Resettlement, relocation, and displacement consist not only of a physical transfer to a new location; a whole a series of changes occur that affect the ways of life of individuals, families, and communities. To paraphrase Robert Gordon (2009: 41), "(Re)settlement involves not only physical movement but also a psychic domain: angst and other anxieties must be allayed for (re)settlers to be settled." Given the complexity of resettlement, it is useful to take to take a human rights-based approach to the issue of resettlement (see Clark, 2009).

There are two major theoretical frameworks dealing with the involuntary resettlement and relocation process. The first of these is one developed by Thayer Scudder and Elizabeth Colson (1982, 2002; see also Scudder 2005, 2009, 2012). Scudder and Colson (1982, 2002) see four general stages relating to projects involving resettlement:

Stage 1. Planning for resettlement (and mitigation) before removal.
Stage 2. Coping with the initial drop in living standards that tends to follow removal.
Stage 3. Initiating economic development and community-formation activities.
Stage 4. Handing over a sustainable resettlement process to the second generation of resettlers and to non-project authority institutions.

With activities that involve resettlement, it is likely that local incomes and living standards for the majority of directly impacted people will decline, at least initially. For some people, over time, livelihoods and well-being may improve, but this is not as common as is impoverishment and decline in living standards.

We know from international experience with large-scale infrastructure projects that compensation alone is insufficient as a means of ensuring the well-being of people affected by development project-related resettlement (Cernea & Maldonado,

© The Author(s) 2023
M. Sapignoli, R. K. Hitchcock, *Anthropology and Ethics*,
https://doi.org/10.1007/978-3-031-39268-9_4

2018; Scudder, 2005; World Commission on Dams, 2000). Resettlement agencies must not only provide people with the means to resettle but also must provide alternative land, and post-resettlement development support (PRDS) and economic opportunities. The development activities must be multi-faceted and culturally relevant. As shown by international experience, gender, age, class, and vulnerability issues also need to be addressed in the course of coming up with sustainable development strategies.

A second major theoretical model dealing with resettlement and risk is that of sociologist Michael Cernea (1995, 1997, 2009) who developed the impoverishment, risks, and reconstruction (IRR) framework. The eight risks of the IRR model are as follows:

1. landlessness
2. joblessness
3. homelessness
4. marginalization
5. food insecurity
6. increased morbidity and mortality
7. loss of access to common property assets
8. social disarticulation

In order to offset these risks and prevent impoverishment from occurring among resettled peoples, efforts must be made (1) to consult fully with those people being relocated, (2) to ensure full local participation in all decisions, (3) work out ways to make people direct beneficiaries, (4) monitor the process carefully, and (5) evaluating the resettlement efforts in light of best international practices (see the documents of the World Bank on indigenous people, resettlement, and participatory development; World Bank, 1995, 2004, 2005, 2017, 2018a, b). These efforts have been made all the more difficult by the watering down of the social safeguards policies of international institutions such as the World Bank in the past several years.

Some of the ways to ensure that people are not affected negatively by conservation and development projects include careful monitoring, the design and implementation of benefit-sharing programs, setting up and running development funds, ensuring the exploitation of natural resources in a sustainable manner, capacity-building of local institutions such as community trusts and village organizations, and the reconstruction of project-affected people's livelihoods at levels that are either equivalent to or better than they were prior to relocation. International best practice in the area of conservation-related resettlement calls for improvement on the livelihoods and well-being of conservation-affected peoples.

Indigenous peoples have all too often faced colonial agencies or international institutions using legal principles to divest them of their land; these come under the "doctrine of discovery", the expropriation principle, the declaration of eminent domain or the defining of the land as *terrus nullus* (empty land) (Daes, 2008; Dahl, 2012). Indigenous people are seeking Free, Prior, and Informed Consent (FPIC) from states and transnational corporations in an effort to become full participants in the planning and decision-making processes related to development

(Alvarado, 2022). A basic legal principle cited by indigenous support organizations involves protections of people from being deprived of their property without just and fair compensation.

Resettlement and relocation are complicated processes, ones that are often extremely hard on the people who are relocated. A major problem with conservation-related resettlement programs is that government officials or agencies tend to focus their attention on the loss of residences (i.e. homes), other buildings (for example, latrines), corrals (livestock pens), and assets such as fruit trees rather than on loss of access to the means of production, especially land, gardens, fields, grazing, and wild resources on which people depend for subsistence and income (Cernea & Maldonado, 2018; Devitt & Hitchcock, 2010). Provision of compensation often works out in such a way that it does not serve as a replacement for lost assets nor a means of ensuring rehabilitation or improvement of livelihoods.

Although the issue of displacement of peoples has been a major subject of discussion internationally for the past several decades, there are relatively few comprehensive legal instruments that deal directly with resettlement. The United Nations has a set of guiding principles (*United Nations Guiding Principles on Internal Displacement*) which have been helpful in providing a set of standards for organizations working with Internally Displaced Persons (IDPs) (Oliver-Smith, 2012). Other organizations have also developed resettlement guidelines, including the Organization of Economic Cooperation and Development (OECD), the Asian Development Bank, the African Development Bank, the Inter-American Development ment Bank, and various non-government organizations (for example, Conservation International, the World Wildlife Fund-US, the Nature Conservancy). Some private mining and oil companies, among others, have guidelines on corporate social responsibility (CSR) which devote some attention to issues of resettlement (Downing, 2002). Issues surrounding corporate social responsibility of transnational corporations and agencies have become significant areas of debate in recent years.

Anthropologists and sociologists have been central to the discussions about improving international resettlement policies and programs (Cernea, 1995, 1997, 2005, 2009; Cernea & Maldonado, 2018; DeWet, 2009, 2012; Oliver-Smith, 2005, 2009a, b, c, 2010; Scudder, 2005, 2009, 2012; Downing & Downing, 2009). In many cases, they have been strong critics of the ways in which development-related, dam-related, and conservation-related resettlement has been handled. Drawing from the lessons of resettlement projects, anthropologists and sociologists have provided useful insights into ways to apply their discipline to policies that affect tens of millions of people around the world.

We believe that it is important to assess carefully arguments relating to the need for conservation-related resettlement. The issue of ethics as they relate to environment and development projects and to human rights has been of growing concern in recent years, as discussed later on in this book. It is crucial to consider the ethical issues involved in the ways in which protected areas are planned, implemented, and managed and to look carefully at the power relations among local people, scientists, park authorities, the state, and international agencies.

More and more emphasis has been placed on cases where 'conservation' has been used as justification for the removals of people from their land and restrictions placed on their natural resource use activities. It is just as important to consider the perspectives of local people along with those of planners, government personnel, non-government organizations, researchers, tourists and other stakeholders involved in conservation and development processes.

Indigenous Peoples and Protected Areas in Central Africa

The Batwa ('Pygmies') of central Africa are well known to the international community and to researchers (Turnbull, 1961, 1983; Hewlett, 2014). Known to some as 'forest peoples' the Batwa of Central Africa are found in a dozen countries, stretching from the Democratic Republic of Congo south to the woodlands of northern and western Zambia. Historically the Batwa of central Africa were tropical and sub-tropical forest hunter-gatherers who interacted to various degrees and in a variety of ways with neighbouring farming peoples (see, for example, Grinker, 1994). Some Batwa are totally dependent on their farmer neighbours; others are semi-dependent, and a number of Batwa groups are autonomous from their neighbours.

It was once thought that tropical forests were the last places other than the high Arctic to be occupied by human populations, presumably because tropical forests lacked high quality starch resources. Tropical forest foraging groups had relatively high residential mobility, moving as frequently as every day or every few days (Binford, 2001; Kelly, 2013). They were or are dependent on a diverse array of resources, but in some cases focused their subsistence on a few key resources. Wild animal meat was traded with nearby farmers in exchange for iron tools, ceramics, and other goods.

The Batwa have been affected substantially by processes of sedentarization, some of which was done at the hands of colonial and post-colonial government authorities. From a social organizational perspective, many Batwa lived in egalitarian, band-level or small-scale societies while their neighbours who tended to belong to middle range societies with a degree of hierarchical differentiation. (for a discussion of small-scale and middle-range societies, see Sapignoli, 2014). Egalitarianism and sharing were valued highly by Batwa and formed an important part of their self-identity.

The African Commission on Human and Peoples' Rights (2005: 87, 2006: 9) stated categorically that it did not intend to give a definitive definition of indigenous peoples. The ACHPR pointed out that there is no global consensus on a universal definition, nor would such a definition be desirable or necessary. Zimbabwe was among the group of African states that asked for clarification of the concept of indigenous peoples in November 2006, prior to the finalization of a draft Declaration on the Rights of Indigenous Peoples (UNDRIP) (African Group of States 2006). The Southern African states (Angola, Botswana, Namibia, South Africa, Zambia, and

Zimbabwe) all voted in favour of the Declaration on the Rights of Indigenous Peoples when it came up for a vote in the United Nations on September 13th, 2007 (Barume, 2009). Several states with Batwa populations did not attend the United Nations session in which the vote was held: C'ote D'ivoire, Equatorial Guinea, Rwanda, and Uganda. One Central African state, Burundi, abstained.

The Batwa have a number of things in common as indigenous peoples. First, they depend heavily on natural resources—forest products—or did in the past. Second, the Batwa have resided for substantial periods in the forests of Central Africa and are generally considered by themselves to be indigenous to the region. Third, the Batwa are highly diverse socially, economically, and politically (Hewlett, 1996; Bahuchet, 2014). Fourth, Batwa have complex interethnic relationships with neighbouring farmers. Fifth, the Batwa are generally at the bottom of the socio-economic as well as political systems of the 12 countries in which they reside. In terms of socio-economic issues, they are marginalized groups while politically they are generally have little power.

There are significant differences among Batwa groups in the degree to which they wish to be autonomous from other groups. There are also differences between neighbouring Batwa groups in the ways in which they earn their livelihoods and employ technology (Turnbull, 1961). In virtually all of the countries of Central Africa in which they are found, the Batwa represent a small minority of the overall population, generally averaging less than 1% of the total (e.g. 0.3% of the Central African Republic, 1.1% of the Democratic Republic of Congo, and slightly less than 1% in Equatorial Guinea and Gabon (Matsuura, 2017).

Some African states have sought to incorporate indigenous peoples into national-level legislation. The Constitution of Cameroon, for example, stipulates that "The State shall ensure the protection of minorities and preserve the rights of indigenous populations in accordance with the law." The term "indigenous" is not defined in the Cameroon Constitution, but Cameroon has created an Indigenous Peoples Development Plan as well as a plan for indigenous and vulnerable peoples in its Poverty Reduction Strategy Paper. Indigenous peoples are also mentioned in legislation in Congo and Burundi. In April 2010, the Central African Republic ratified Convention 169 of the ILO, the first African country to do so.

The Batwa peoples exhibit both similarities and differences in their goals and objectives. They all wish to have their human rights respected. They want ownership and control over their own land and natural resources; and they want the right to be to participate through their own institutions in the political process at the national, regional, and international levels. The Batwa all want to ensure that they and their children are protected in the face of conflicts and efforts to dispossess them or deprive them of their rights to life, liberty, and livelihoods. They all hope to pursue and preserve their own cultural traditions, and they wish for their children to learn about their own cultures and to be allowed to speak mother-tongue languages in schools.

Batwa populations are found currently in 12 central African countries: Angola, Burundi, Cameroon, Central African Republic, Congo, the Democratic Republic of Congo, Equatorial Guinea, Gabon, Malawi, Rwanda, and Uganda (see Table 4.1).

Table 4.1 Data on Batwa Populations in Central Africa

Country	Population Size (July, 2022 estimate)	Size of Country (square kilometres)	Numbers of Batwa (National)
Angola	34,795,287	1,246,700	Batwa 1000
Burundi	12,696,478	27,830	Batwa 70,000 (with Rwanda)
Cameroon	29,321,637	473,440	Bedzan 400, BaKola 4000; Baka 40,000
Central African Republic (CAR)	5,454,333	622,984	Baka 30,000–40,000 (with Congo, Cameroon), Aka (Bayaka) 30,000 (with Congo)
Congo Republic (RC)	5,546,307	342,000	Baka 70,000 (with Congo, Cameroon), Bakoya 2600 (with Gabon), Aka 30,000 (with CAR)
Democratic Republic of Congo (DRC)	105,407,721	2,344,858	Asua (Mbuti) 10,000, Efe 10,000, Batwa 6000, Barhwa (Kivu Twa) 6000, Luba Cwa 2000, Batembo, 4000, Basua, 26,000
Equatorial Guinea	1,679,172	28,051	Batwa, 3000
Gabon	2,340,613	267,667	Bakoya 2600 (with Congo), Babongo 3000
Malawi	20,794,353	118,484	500 were in Chongoni Forest reserve
Rwanda	13,173,730	26,388	Batwa (Twa) 70,000 (with Burundi)
Uganda	46,205,983	241,038	Batwa 2000
Zambia	19,642,123	752,614	Batwa 1000
Totals	People in 12 countries	6,492,054 km^2	Ca. 400,000 Batwa

Note: Data obtained from fieldwork, interviews of researchers and non-government organization personnel, and from *The World Factbook* (2023); *The Indigenous World* (2022, 2023), Bahuchet (2014: 8, Table 1.1), and Olivero et al. (2016)

Known historically as 'tropical forest peoples' the Batwa have existed as specialized hunters and gatherers and traders who interact extensively with other peoples, many of whom they were dependent upon for access to food, manufactured goods and services. Inter-ethnic relations between Batwa and farmers were very important for both sets of groups (see Rupp, 2011). Government policies toward Batwa in Central Africa vary (Hewlett, 1996, 2000) as do the perceptions of the Batwa as people.

Some of the features of Batwa populations in these countries include high percentages of people living below the poverty line, low rates of educational attainment and relatively low literacy rates, high rates of unemployment, largely rurally oriented economies, low degrees of access to land and resources, low to moderate health statuses, high rates of exposure to discriminatory practices at the hands of other groups and the state (Hewlett, 2014).

An examination of the various interactions between indigenous peoples and the nation-states of Central Africa reveals that there is a wide range of variation in the ways in which indigenous groups are treated. In the case of Burundi, for example,

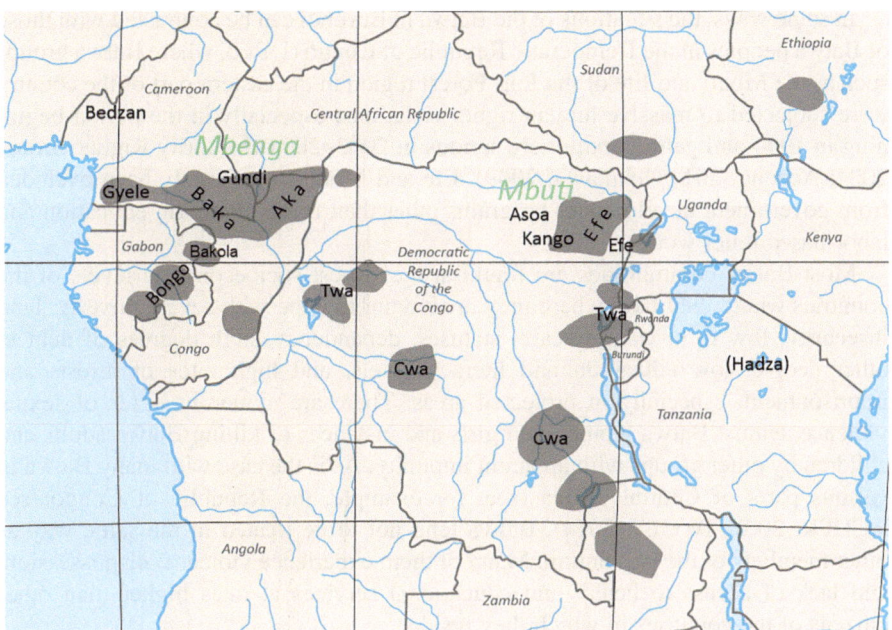

Fig. 4.1 Batwa distribution in Central Africa. Source: Jerome Lewis, University College London

which has experienced cycles of violence and clashes among various groups over time, the situation of some members of the minority Batwa population, which makes up some 1–2% of the total population of the country of some 12.5 million, has improved over time. Burundi has signed and ratified a number of the international conventions that deal with human rights, including the *African Charter on Human and Peoples' Rights* and the *Convention on the Rights of the Child* (Barume, 2014). The 2004 Constitution of Burundi guarantees seats in both the Parliament and the Senate for members of the Batwa. Both government and civil society in Burundi are seeking to promote Batwa rights and to provide development opportunities and access to health and education services (Fig. 4.1).

The Burundi government claims to be fully aware of the marginalized situations of the Batwa (ACHPR, 2007: 16–27). There are, however, some government officials who feel that the Batwa are attempting to exclude themselves from the rest of the country by wishing to be identified as indigenous (ACHPR, 2007: 26–27). A major problem facing the Batwa in Burundi is debt bondage; several thousand Batwa are in complex positions *vis a vis* people to whom they owe money and for whom they have to work under difficult and slavery-like conditions. It is interesting to note in this regard that the government of Burundi is one of the few African countries that has yet to ratify the *Supplementary Convention on the Abolition of Slavery, the Slave Trade, and Institutions and Practices Similar to Slavery* (ACHPR, 2007: 16).

In some ways, the situations of the Batwa in Burundi can be contrasted with those of Batwa peoples in the Democratic Republic of Congo (DRC), where Batwa groups such as the Mbuti and Efe of the Ituri Forest region in the eastern part of the country were subjected to massive human rights violations, especially in the period beginning in 1994 and getting extremely serious in 2003–2004 (Minority Rights Group, 2004; Acheng, 2011; Ichikawa, 2009). Efe and Mbuti have largely been excluded from government development programs other than those involving education and labor-based relief work.

Most Batwa communities are found at the lowest socioeconomic levels of the countries where they live, where they are having to cope with severe poverty, land insecurity, low to moderate health statuses, dependency, high degrees of debt to other people, low education and literacy levels, and high rates of arrests and imprisonment, especially in protected areas. There are numerous cases of sexual violence against Batwa women and girls and instances of killing Batwa adults and children by other groups with apparent impunity. As is the case with many Batwa in various parts of Central Africa (see, for example, the Republic of Congo; see ACHPR, 2007; IWGIA, 2014), Batwa tend not to be treated in the same way as other members of the population. Many of them experience violence, dispossession, and lack of access to health and educational services at rates higher than other citizens of the countries in which they reside.

A major issue facing the Batwa of Central Africa has been the involuntary relocation out of national parks and other protected areas (Schmidt-Soltau, 2003; Schmidt-Soltau & Brockington, 2007; Cernea & Schmidt-Soltau, 2006; Tranquilli et al., 2014). Table 4.2 presents data on these evictions.

Conservationists reacted negatively to the reports on the involuntary relocation of people from Central African protected areas (see, for example Redford & Fearn, 2007; Webber & Vedder, 2002). Anthropologists and indigenous support organizations took the position that the relocation processes were having serious impacts on indigenous people.

It is useful to provide some examples of where these protected areas have played significant roles in resettlement. In the case of what became the Kibale Forest Reserve (now Kibale National Park) in Uganda, some 33,000 people were removed at gunpoint by the Ugandan Army and the Uganda Forestry Department in 1991 and 1992 in order to facilitate the establishment of Kibale National Park in 1993. Part of the justification for the removals was to prevent what the Ugandan government described as 'overutilization of resources' and 'habitat destruction' by residents of the area (Marquardt, 1994; Feeney, 1998; Naughton-Treves, 1999; Naughton-Treves & Sanderson, 1995; Naughton-Treves et al., 2005). Some of the people removed from the Kibale area were Abandyala Batwa who had been there for generations (Wily & Kabananukye, 1996). It is important to note, however, that the Ugandan Constitutional Court ruled on 19 August 2021 that compensation should be paid to those Batwa in the country who were forcibly relocated out of protected areas (Kasule, 2021; www.minorityrights.org, accessed 26 February 2023). The Constitutional Court decision arises from a 2013 petition filed by the

Table 4.2 National Parks, Game Reserves, and Conservation Areas in Central Africa That Resulted in the Involuntary Resettlement of Local Batwa Populations

Park or Reserve Area, Date of Establishment, Size in km^2	Country	Comments
Dja Biosphere Reserve, 1950, 5260 km^2	Cameroon	Baka traditional hunting is allowed, agriculture is not allowed, 7800 Baka
Korup National Park, 1986, 1259 km^2	Cameroon	1465 resettled, some compensation paid, 5 villages inside the park, 1 village resettled in 2000
Kale Lobeke National Park, 2001, 2180 km^2	Cameroon	4000 relocated, some use zones inside the park, Baka sidelined
Dzanga-Ndoki National Park, 1990, 1220 km^2	Central African Republic	350 Batwa relocated
Nouabale'Ndoki National Park, 1993, 3865 km^2	Republic of Congo	3000 people, WWF considers area uninhabited, Bangoma Batwa hunt in zones outside of the park
Odzala National Park, 1935, 13,000 km^2	Republic of Congo	9800 people, villages along margins, Odzala Foundation, African Parks and Congo government cooperation
Sanga Transnational Site (TNS), 2012, 178,795 km^2	Cameroon, CAR, Republic of Congo	18,000 Baka, Bangando, Bakwele, Sanga-Sanga in buffer zones, use zones in Lobeke but foraging not permitted in CAR or Republic of Congo
Kahuzi-Biega National Park, 1970, 6000 km^2, World Heritage Site, 1980	Democratic Republic of Congo	6000 people dispossessed, habitation prohibited, Batwa live on park margins
Bwindi Impenetrable National Park, 1991, 331 km^2, World Heritage Site, 1994	Uganda	50–100 Batwa families resettled and marginalized, violent attacks in March 1999
Kibale National Park, 1993, 795 km^2	Uganda	33,000 people were evicted in 1991, 1992, and 1993
Rwenzoris National Park, 1991, 1998, World Heritage Site, 1994 996 km^2	Uganda	Batwa resettled in the 1980s, Bakonjo had benefits from the national park

Note: Data obtained from field work, the IUCN/World Conservation Union, the World Heritage Site registry

United Organisation for Batwa Development in Uganda against the Attorney General, Uganda Wildlife Authority and National Forestry Authority (Kasule, 2021: 1).

In another case in Cameroon, Worldwide Fund for Nature (WWF) was investigated for having some of its park guards engage in actions that violated the human rights of local Batwa communities (Barber, 2022). Similar investigations were carried out in the case of the Republic of Congo, where the Integrated and Transboundary Conservation of Biodiversity in the Basins of the Republic of Congo, TRIDOM II project was being implemented by the Worldwide Fund for Nature and the government of Congo, resulting in the suspension of the project in March 2019. This suspension occurred after complaints were made by six indigenous Baka communities located in the vicinity of the proposed Messok-Dja

Protected Area in the northern Republic of Congo regarding severe mistreatment. These coercive approaches have been condemned widely and have led to the stopping of international financial support for WWF projects in Cameroon, the Central African Republic, the Democratic Republic of Congo, and the Republic of Congo. These cases have raised international awareness of some of the complexities that exist among conservation organizations vis a vis their treatment of indigenous peoples, an issue raised nearly two decades ago by anthropologist Mac Chapin (Chapin, 2004).

The African Commission on Human and Peoples Rights Working Group on Indigenous Populations/Communities and the Special Rapporteur on the human rights and fundamental freedoms of indigenous people have undertaken trips to a number of Central African countries and has highlighted some of the issues facing indigenous peoples and made recommendations aimed at the improvement of their statuses and well-being (e.g. Burundi, the Congo Republic, the Democratic Republic of Congo, Gabon, Rwanda, and Uganda).

On 11 February 2016, Survival International filed a human rights complaint alleging the World Wildlife Fund's complicity in the torture and abuse of Baka hunters in Cameroon's national parks. Beginning in the mid-90s, the Baka were evicted from their traditional hunting territories so that these lands could be set aside for conservation. The Baka were provided with no meaningful alternative ways to feed their families. Cameroonian and Western human rights advocates, many of whom have worked for over a decade to bring this issue to public attention. Survival International dropped this lawsuit in 2018. WWF continued to use militant tactics that violate Baka rights in the latter part of 2019 and 2020 and as a result lost their contract to do the work in Cameroon and the Democratic Republic of Congo.

In March 2019 WWF—Worldwide Fund for Nature—commissioned an independent panel of experts to review human rights allegations levelled at government park rangers in areas where WWF works in Cameroon, the Central African Republic, the Republic of Congo, the Democratic Republic of the Congo (DRC), India, and Nepal. WWF maintained that it takes all allegations of human rights abuse seriously. WWF and other organizations with whom it was working had their funds cut off by the United Nations Development Program because of the human rights violations that had occurred (Minority Rights Group International, 2020). Another problem has been the intimidation and murders of environmental rights defenders in Central Africa and other parts of the world (Global Witness, wwwl.globalwitness.org, accessed 15 March 2023).

The Old Way and the New Way

It is possible to compare traditional and modern strategies of what can be termed 'the Old Way' and the 'New Way' (Table 4.3).

Central African Batwa and southern African San exhibit similar principles in the ways in which they handle they environment. They both seek to respect nature and to

Table 4.3 Comparison of 'The Old Way' and 'The New Way' with respect to Indigenous peoples in central and southern Africa

The Old Way	The New Way
The old rules	The new rules
Living entirely from natural resources	Living from domestic resources
Hunter-gatherers (foragers)	Farmers, pastoralists, and industrial systems
No domestic animals	Domestic animals
No agriculture	Agriculture
Self-sufficient	Dependency on others
Savanna and forest living	Living in diverse environments
Seasonality important	Much buffering against seasonal variability
Use of diversity of resources	Use of a more restricted number and type of resources
Simple technology	Complex technology
Subsistence hunting of importance	Subsistence hunting of limited importance
No exchange with non-foragers	Substantial exchange with non-foragers
Sharing and reciprocity	Exchange based on monetary value
Egalitarianism	Non-egalitarian systems
Gender equity	Gender inequality
Conflict-resolution mechanisms include talking out problems	Conflict resolution mechanisms are complex including relying on state systems
Consensus-based decision making	Hierarchical decision-making
Run down animals	Use of complex technology to obtain animals
Every species is challenged to push to the extremes	Variability in adaptations and adjustments
Co-existence with predators	No co-existence with predators except in zoos
Treat elephants with respect	Elephants treated in complex, sometimes exploitative ways
Fire is the main way that the environment is manipulated	Fire is discouraged except insofar as it is used to manage ecosystems and burn off wastes
Using traditional technology	Using modern technology including high-powered weapons
Conservation as a major principle	Exploitation as a major principle, some conservation also practiced
Stability	Instability

Note: Data obtained from Marshall Thomas (1990, 2006, personal communications, 2014–2023)

ensure that they employ sustainable strategies in the ways in which they manage their lands. Admittedly, there are cases where indigenous peoples over-exploit their resources, as can be seen in cases where wild animals and plants are exploited beyond their carrying capacity. Conservationists have remarked that there are cases where indigenous peoples have exploited sizable numbers of resources, to the point where the exploitation rates are unsustainable. While admitting that this is sometimes the case with resource exploitation, especially in cases where wild species are being sold to generate income, indigenous peoples argue that they are conservation-minded and that the methods they use are aimed at keeping exploitation rates as low as possible.

Free, Prior, and Informed Consent (FPIC)

The concept of free, prior, and informed consent is deeply intertwined with the rights of indigenous people. FPIC is a specific right that pertains to indigenous people and is recognized in the UNDRIP (Article 31). It allows indigenous people to give or withhold consent to a project that may affect them or their territories. Once they have given their consent, they can withdraw it at any stage. Furthermore, FPIC enables them to negotiate the conditions under which the project will be designed, implemented, monitored, and evaluated. Under current policies of the United Nations Development Program, the World Bank, the African and Asian Development ment Banks, the International Finance Corporation, and the Food and Agriculture Organization of the United Nations, Free, Prior and Informed consent is mandated to be followed in any projects involving social and environmental impact assessments and social safeguards policies.

FPIC is one of the most important principles that indigenous peoples have in order to protect their human rights (United Nations Commission on Human Rights, 2005; Ward, 2011; Tonkinson, 2017; Doyle et al., 2019; Iseli, 2020). Under the principles of FPIC it is the duty of the state and private companies to obtain free, prior, and informed consent prior to the initiation of any project. FPIC has its origins in the concept of 'native title' from common law. It is enshrined in the International Protocol on Civil and Political Rights (IPCPR) and in some state legislation, such as that of the Philippines Indigenous Peoples Rights Act of 1997. FPIC allows for constructive engagements with indigenous peoples. It is the core international standard that allows indigenous rights to be fully realized.

The critical issues involved with FPIC are (1) enforcement, and (2) compliance. There are relatively few cases worldwide where FPIC has been observed fully. In many cases, states and companies tend to emphasize free, prior, and informed *consultation*, leaving aside consent. This is how the World Bank views FPIC—as consultation, not allowing local people to give consent on a project (De Moerloose, 2020). Corporate Social Responsibility (CSR) of international businesses calls for FPIC, but in general there is little enforcement of FPIC regulations. There are cases, however, where Chief Executive Officers of companies such as Rio Tinto were forced to resign after failing to take into consideration indigenous groups' statements about the need for protection of their sacred sites, as occurred in 2020 in Australia when two ancient sacred sites were destroyed by the mining company (Mercer, 2020). The destruction helped fuel the push for a new Aboriginal Cultural Heritage Bill 2020 in Australia.

Free, Prior and Informed Consent has become a customary international human rights norm. One of the problems is that the United Nations Declaration on the Rights of Indigenous People, which incorporates FPIC is a declaration and not a convention, which does not have the power of international law. At a minimum FPIC calls for the consultation of affected peoples when any development project is being considered. There is an enormous gap between norms that are being developed within human rights jurisprudence and state practice.

Free Prior and Informed Consent has a number of important principles. Special attention has to be paid to the political, economic and social context of an FPIC process to ensure that it is fully free from coercion or manipulation (e.g. through giving of bribes to local leaders). For an FPIC process to be done appropriately, it must start prior to any efforts to plan and implement a project. In terms of information dissemination, all peoples in a potential project area must have access to and share accurate information pertaining to the plans for the potential impacts of a project. What this means, in effect, is that local people who are potentially project-affected require technically accurate social and environmental impact assessments and statements. These must be presented in a variety of forms: oral presentations, handouts, information sheets that are easily understandable and in mother tongue languages, and posters presenting data in local languages. Meetings with local people must be held prior to any planning of a project. States and companies must ensure that this is done in line with customary methods of decision-making. Project-affected peoples must have an impact on the final decision about a project. They can say yes, no, or yes but with conditions. The meetings must be done in such a way that all sub-groups of a community are consulted. Gender, age, and physical conditions of participants must be taken into consideration. Consultation must be accompanied with consent, in other words, indigenous and other people have veto power over projects. To summarize, free means no element of coercion, prior means done ahead of time, informed means all information must be available to affected people, and consent means that local people have the right, after considering a project, to allow collectively for the project to go ahead or to be terminated.

As the Food and Agriculture Organization (2015) notes, all elements within FPIC are interlinked, and they should not be treated as separate elements. FPIC refers to a process that is self-directed by the community from whom consent is being sought, with no elements or coercion or pressure to come to a specific decision. Meetings and discussions should be held at times and places convenient to the people involved. There should be no time limitations placed on the process, and it is crucial that the process begins at the community level at the same time or before decisions are being made by the state or company involved. Careful monitoring and evaluation (M&E) must be done throughout the process. Transparency and accountability are key principles in the FPIC process. Cost-benefit analyses must be carried out during the course of project design and implementation, and all information on risks, costs, and benefits must be provided to the communities. All the information must be made available not only to community leaders but to women, youth, the elderly, and persons with disabilities. Consent must be obtained collectively, and the decisions conveyed to the highest levels of government and companies involved in the FPIC process.

FPIC is a participation right, and one that is tied directly to self-determination of a people or set of communities. FPIC is a tool used to ensure that indigenous and other people's priorities are taken into full account. It is important to note that in the case of removals of people from the various central African national parks and protected areas, in no cases were people informed ahead of time, they were not given full information, and they had no opportunity to say whether or not they agreed with the decision of the states about their future. Local people were not given the opportunity to discuss the projects ahead of time, and they were often moved before they could organize themselves to oppose the decision. These strategies made it difficult for them to seek legal counsel and to take the state to court for their actions.

Environment and Social Impacts of Conservation-Related Resettlement: Some Conclusions

Judging from data on conservation-related resettlement in Central Africa, a whole set of impacts of these processes can be discerned. These include the following:

- People are generally worse off economically and socially after they are resettled.
- Impoverishment is common after resettlement.
- Resettled people find themselves in situations where their social and political systems have deteriorated.
- Health conditions deteriorate, with greater problems of nutritional and physical stress; morbidity (illness) rates often increase
- Resettled people tend to be worse off as a result of increased diseases, including ones that are water-borne such as cholera and typhoid and there are more problems with pandemic diseases such as HIV AIDS, tuberculosis, and COVID 19
- There are greater problems of inter and intra-community competition and conflict
- Environmental deterioration in resettlement areas is common
- There are greater problems with landlessness or inadequate access to land in resettlement areas
- Levels of violence increase after resettlement, including gender-based violence, domestic abuse, and child abuse; there is also greater inter-group violence
- Community-based organizations tend to deteriorate, and they have to reorganize in order to remain viable.
- Increased stratification occurs among resettled people over time

Judging from the experience of the 12 African countries that support Batwa and other indigenous people, strategies that have assisted them people include incorporating reference to indigenous people in national constitutions, recognizing the rights of indigenous people in national legislation, ensuring that there is indigenous representation in national parliaments, providing targeted support programs for indigenous people, mapping of indigenous lands, ensuring security of tenure over indigenous land, and setting up specific programs to assist indigenous people who have been resettled including ones that incorporate just and fair compensation. According to indigenous peoples in central Africa international non-government organizations and international organizations including the World Bank and the United Nations Development Program need to increase their social safeguards in order to protect both human rights and biodiversity.

References

Acheng, I. (2011). *Batwa of the Democratic Republic of Congo*. International Work Group for Indigenous Affairs.

African Commission on Human and Peoples' Rights. (2005). *Report of the African Commission's Working Group of Experts on Indigenous Populations/Communities*. African Commission on Human and People's Rights and International Work Group for Indigenous Affairs.

African Commission on Human and Peoples' Rights. (2006). *Indigenous peoples in Africa: The forgotten peoples? The African Commission's Work on Indigenous Peoples in Africa*. African Commission on Human and People's Rights and International Work Group for Indigenous Affairs.

African Commission on Human and Peoples' Rights, Working Group on Indigenous Populations/ Communities in Africa. (2007). *Research and Information Visit to the Republic of Burundi, 27 March–9 April, 2005*. African Commission on Human and Peoples' Rights, African Union and International Work Group for Indigenous Affairs.

Alvarado, L. J. (2022). *Study on consultation and free, prior and informed consent with indigenous peoples in Africa*. International Work Group for Indigenous Affairs.

Bahuchet, S. (2014). Cultural diversity of African pygmies. In B. S. Hewlett (Ed.), *Hunter-gatherers of The Congo Basin: Cultures, histories, and biology of African pygmies* (pp. 1–29). Transaction Publishers.

Barber, N. J. (2022). *Baka representation: Rights, videomaking, and indigenous identity in South-eastern Cameroon*. Ph.D. Dissertation, McGill University, Montreal, Ontario, Canada.

Barume, A. (2009). Responding to the concerns of the African states. In C. Charters & R. Stavenhagen (Eds.), *Making the declaration work: The United Nations declaration on the rights of indigenous peoples* (pp. 170–182). International Work Group for Indigenous Affairs.

Barume, A. K. (2014). *The land rights of indigenous peoples in Africa, with specific focus on central, eastern, and southern Africa*. IWGIA document 128. Revised and updated 2014. International Work Group for Indigenous Affairs.

Binford, L. R. (2001). *Constructing frames of reference: An analytical method for archaeological theory building using ethnographic and environmental data sets*. University of California Press.

Cernea, M. M. (1995). Understanding and preventing impoverishment from displacement: Reflections from the state of knowledge. *Journal of Refugee Studies, 8*(3), 245–264.

Cernea, M. M. (1997). The risks and reconstruction model for resettling displaced populations. *World Development, 25*(10), 1569–1587.

Cernea, M. M. (2005). Restriction of access is displacement: A broader concept and policy. *Forced Migration Review, 23*, 48–49.

Cernea, M. M. (2009). Financing for development: Benefit-sharing mechanisms in population resettlement. In A. Oliver-Smith (Ed.), *Development and dispossession: The crisis of forced displacement and resettlement* (pp. 49–76). School for Advanced Research.

Cernea, M., & Schmidt-Soltau, K. (2006). Poverty risks and National Parks: Policy issues in conservation and resettlement. *World Development, 34*(10), 1808–1830.

Cernea, M. M., & Maldonado, J. K. (Eds.). (2018). *Challenging the prevailing paradigm of displacement and resettlement: Risks, impoverishment, legacies, solutions*. Routledge.

Chapin, M. (2004). A challenge to conservationists. *World Watch Magazine, 11–12*, 17–31.

Clark, D. (2009). Power to the people: Moving towards a rights-respecting resettlement framework. In A. Oliver-Smith (Ed.), *Development and dispossession: The crisis of forced displacement and resettlement* (pp. 181–199). School for Advanced Research.

Daes, E.-I. (2008). *Indigenous peoples: Keepers of our past, custodians of our future*. International Work Group for Indigenous Affairs.

Dahl, J. (2012). *The indigenous space and marginalised peoples in the United Nations*. Palgrave Macmillan.

De Moerloose, S. (2020). *World Bank environmental and social conditionality as a vector of sustainable development*. Schulthess éditions romandes.

Devitt, P., & Hitchcock, R. K. (2010). Who drives resettlement? The case of Lesotho's Mohale Dam. *African Study Monographs, 31*(2), 57–106.

DeWet, C. (2009). Does development displace ethics? The challenge of forced resettlement. In A. Oliver-Smith (Ed.), *Development and dispossession: The crisis of forced displacement and resettlement* (pp. 77–96). School for Advanced Research.

DeWet, C. (2012). The application of international resettlement policy in African Villagization projects. *Human Organization, 71*(4), 395–406.

Downing, T. (2002). *Avoiding new poverty: Mining-induced displacement and resettlement*. International Institute for Environment and Development.

Downing, T. E., & Garcia-Downing, C. (2009). Routine and dissonant cultures: A theory about the psych-sociocultural disruptions of involuntary displacement and ways to mitigate them without inflicting more damage. In A. Oliver-Smith (Ed.), *Development and dispossession: The crisis of forced displacement and resettlement* (pp. 233–255). School for Advanced Research.

Doyle, C., Whitmore, A., & Tugendhat, H. (Eds.). (2019). *Free prior informed consent protocols as instruments of autonomy: Laying foundations for rights based engagement*. European Network on Indigenous Peoples (ENIP) and Infoe.

Feeney, P. (1998). *Accountable aid: Local participation in major projects*. Oxfam.

Food and Agriculture Organization of the United Nations. (2015). *Free, prior, and informed consent: An Indigenous Peoples' right and a good practice for local communities*. Food and Agriculture Organization of the United Nations.

Gordon, R. J. (2009). Hiding in full view: The "Forgotten" Bushman Genocides in Namibia. *Genocide Studies and Prevention, 4*(1), 29–57.

Grinker, R. R. (1994). *Houses in the rainforest: Ethnicity and inequality among farmers and foragers in Central Africa*. University of California Press.

Hewlett, B. S. (1996). Cultural diversity among African pygmies. In S. Kent (Ed.), *Cultural diversity among twentieth century foragers: An African perspective* (pp. 215–244). Cambridge University Press.

Hewlett, B. S. (2000). Central African Government's and international NGOs' perceptions of Baka pygmy development. In P. P. Schweitzer, M. Biesele, & R. K. Hitchcock (Eds.), *Hunters and gatherers in the modern world: Conflict, resistance, and self-determination* (pp. 380–390). Berghahn Books.

Hewlett, B. S. (Ed.). (2014). *Hunter-gatherers of The Congo Basin: Cultures, histories, and biology of African pygmies*. Transaction Publishers.

Ichikawa, M. (2009). Forests and indigenous people in post-conflict democratic Republic of Congo. In G. Kato & A. Uyar (Eds.), *Question of poverty and development in conflict and conflict resolution* (Afrasian Studies) (pp. 211–226). Ryukoku University.

Iseli, C. (2020). The operationalization of the principle of free, prior and informed consent: A duty to obtain consent or simply a duty to consult? *UCLA Journal of Environmental Law and Policy, 38*(2), 259–275.

IWGIA (International Work Group for Indigenous Affairs). (2014). *Country technical notes on indigenous peoples issues: The Republic of Congo*. International Work Group for Indigenous Affairs (IWGIA).

Kasule, F. (2021). Compensate Batwa – Court. *New Vision*, 24 August 2021.

Kelly, R. L. (2013). *The lifeways of hunter-gatherers: The foraging Spectrum*. Cambridge University Press.

Marquardt, M. (1994). Settlement and resettlement: Experience from Uganda's national parks and reserves. In C. C. Cook (Eds.), *Involuntary resettlement in Africa* (pp. 147–149). World Bank Technical Paper 227. The World Bank.

Marshall Thomas, E. (1990). The old way. *The New Yorker*, October 15, 1990.

Marshall Thomas, E. (2006). *The old way: A story of the first people*. Farrar, Straus, Giroux.

Matsuura, N. (2017). Humanitarian assistance from the viewpoint of hunter-gatherer studies: Cases of Central African Forest Foragers. *African Study Monographs, Supplementary Issue, 53*, 117–129.

Mercer, P. (2020). *Aboriginal rock art destroyed by mining company*. Report to the Australian Broadcasting Corporation.

Minority Rights Group International. (2004). *'Erasing the Board': Report of the International Research Mission into Crimes under International Law Committed Against the Bambuti Pygmies in the Eastern Democratic Republic of Congo*. Minority Rights Group International.

Minority Rights Group International. (2020). *Violent conservation: WWF's failure to prevent, respond to, and remedy human rights abuses committed on its watch*. Minority Rights Group International.

Naughton-Treves, L. (1999). Whose animals? A history of property rights to wildlife in Toro, Western Uganda. *Land Degradation and Development, 10*, 311–328.

Naughton-Treves, L., Holland, M. B., & Brandon, K. (2005). The role of protected areas in conserving biodiversity and sustaining local livelihoods. *Annual Review of Environment and Resources, 30*, 219–252.

Naughton-Treves, I., & Sanderson, S. E. (1995). Property, politics, and wildlife conservation. *World Development, 23*, 1265–1275.

Olivero, J., Fa, J. E., Farfán, M. A., Lewis, J., Hewlett, B., Breuer, T., Carpaneto, G. M., Fernández, M., Germi, F., Hattori, S., Head, J., Ichikawa, M., Kitanaishi, K., Knights, J., Matsuura, N., Migliano, A., Nese, B., Noss, A., Ekoumou, D. O., Paulin, P., Real, R., Riddell, M., Stevenson, E. G. J., Mikako Toda, J., Vargas, M., Yasuoka, H., & Nas, R. (2016). Distribution and numbers of pygmies in central African forests. *PLoS One, 11*(1), e0144499. https://doi.org/10.1371/journal.pone.0144499

Oliver-Smith, A. (2005). Applied anthropology and development-induced displacement and resettlement. In S. Kedia & J. John van Willigen (Eds.), *Applied anthropology: Domains of application* (pp. 189–219). Praeger Publishers.

Oliver-Smith, A. (2009a). Introduction: Development-forced displacement and resettlement: A global human rights crisis. In A. Oliver-Smith (Ed.), *Development and dispossession: The crisis of forced displacement and resettlement* (pp. 3–23). School for Advanced Research.

Oliver-Smith, A. (2009b). Evicted from Eden: Conservation and the displacement of indigenous and traditional peoples. In A. Oliver-Smith (Ed.), *Development and dispossession: The crisis of forced displacement and resettlement* (pp. 141–162). School for Advanced Research.

Oliver-Smith, A. (Ed.). (2009c). *Development and dispossession: The crisis of forced displacement and resettlement*. School for Advanced Research.

Oliver-Smith, A. (2010). *Defying displacement: Grassroots resistance and the critique of development*. University of Texas Press.

Oliver-Smith, A. (2012). Climate change, displacement and resettlement: The need for human rights based public policy. *Anthropology Newsletter, 53*(10), 25.

Redford, K., & Fearn, E. (Eds.). (2007). *Protected areas and human displacement: A conservation perspective* (Working paper no. 29). Wildlife Conservation Society.

Rupp, S. (2011). *Forests of belonging: Identities, ethnicities, and stereotypes in the Congo River Basin*. University of Washington Press.

Sapignoli, M. (2014). Mobility, land use, and leadership in small-scale and middle-range societies. *Reviews in Anthropology, 43*, 1–44.

Schmidt-Soltau, K. (2003). Conservation–related resettlement in Central Africa: Environmental and social risks. *Development and Change, 34*(3), 525–551.

Schmidt-Soltau, K., & Brockington, D. (2007). Protected areas and resettlement: What scope for voluntary relocation? *World Development, 35*(12), 2182–2202.

Scudder, T., & Colson, E. (1982). From welfare to development: A conceptual framework for the analysis of dislocated people. In A. Hansen & A. Oliver-Smith (Eds.), *Involuntary migration and resettlement: The problems and responses of dislocated people* (pp. 267–287). Westview Press.

Scudder, T., & Colson, E. (2002). Long-term research in the Gwembe Valley, Zambia. In R. V. Kemper & A. P. Royce (Eds.), *Chronicling cultures: Long-term field research in social anthropology* (pp. 197–238). AltaMira Press.

Scudder, T. (2005). *The future of large dams: Dealing with social, environmental, institutional, and political costs*. James and James Science Publishers.

Scudder, T. (2009). Resettlement theory and the Kariba case: An anthropology of resettlement. In A. Oliver-Smith (Ed.), *Development and dispossession: The crisis of forced displacement and resettlement* (pp. 25–47). School for Advanced Research.

Scudder, T. (2012). Resettlement outcomes of large dams. In C. Tortajada, D. Altinbelek, & A. K. Biswas (Eds.), *Impacts of large dams: A global assessment* (pp. 37–68). Springer.

Scudder, T., & Colson, E. (1982). From welfare to development: A conceptual framework for the analysis of dislocated people. In A. Oliver-Smith, A. Hansen, & A. (Eds.), *Involuntary migration and resettlement: The problems and responses of dislocated people* (pp. 267–287). Westview Press.

Tranquilli, S., Abedi-Lartey, M., Abernethy, K., Amsini, F., Asamoah, A., et al. (2014). Protected areas in tropical Africa: Assessing threats and conservation activities. *PLoS One, 9*(12), e114154. https://doi.org/10.1371/journal.pone.0114154

Turnbull, C. M. (1961). *The forest people: A study of the pygmies of The Congo*. Simon and Schuster.

Turnbull, C. M. (1983). *The Mbuti pygmies: Change and adaptation*. Holt, Rinehart and Winston.

United Nations. (2007). *United Nations declaration on the rights of indigenous peoples (UNDRIP)*. United Nations general assembly resolution 61/295, September 13th, 2007. United Nations.

U.N. Commission on Human Rights, Sub-Committee on the Promotion and Protection of Human Rights Working Group on Indigenous Populations. (2005). *Working Paper: Standard-Setting: Legal Commentary on the Concept of Free, Prior and Informed Consent*, ¶ 57, U.N. Doc. E/CN.4/Sub.2/AC.4/2005/WP.1, 2005 prepared by Antoanella-Iulia Motoc and the Tebtebba Foundation).

Ward, T. (2011). The right to free, prior, and informed consent: Indigenous peoples' participation rights within international law. *Northwestern Journal of International Human Rights, 10*(2), 54–84.

Webber, B., & Vedder, A. (2002). *In the kingdom of gorillas: The quest to save Rwanda's mountain gorillas*. Simon and Schuster.

Wily, L., & Kabananukye, K. (1996). Pygmies (Abayanda) of South Western Uganda: Reaching the End of the Road – and the Beginning. *Indigenous Affairs, 1996*(4), 26–34.

World Bank. (1995). *World Bank participation sourcebook*. Environment Department Papers 019. World Bank.

World Bank. (2004). *Involuntary resettlement sourcebook: Planning and implementation in development projects*. World Bank.

World Bank (2005) Indigenous Peoples. In The World Bank operational manual, operational policies 4.10, World Bank, ed. Pp. 1–13. : World Bank.

World Bank. (2017). *The World Bank environmental and social framework*. World Bank.

World Bank. (2018a). *Guidance note for borrowers – Environment and social framework for IFP operations: ESS5: Land acquisition, restrictions on land use and involuntary resettlement*. The World Bank.

World Bank. (2018b). *Guidance note – ESS7: Indigenous peoples/sub-Saharan African historically underserved traditional local communities*. The World Bank.

World Commission on Dams. (2000). *Dams and development. A new framework for decision-making*. Earthscan.

Chapter 5
Indigenous Peoples' Strategies for Coping with Protected Area Policies and Treatment

Introduction

In this chapter we consider some of the ways that indigenous peoples have attempted to cope with protected area policies of the nation-states in which they reside. We begin with a discussion of the countries of South America, some of which have pursued forward-thinking policies toward indigenous people. Some South American countries have set aside indigenous protected areas, while others, such as Argentina, Brazil, Paraguay, and Peru, have engaged in efforts to map indigenous lands and demarcate them officially (Poole, 1995; Chapin et al., 2005; Chapin & Threlkeld, 2001; Herlihy & Knapp, 2003). Indigenous peoples, for their part, have engaged in demonstrations for their rights at the local, national, and international levels, as occurred, for example, at the United Nations Conference on Environment and Development in Rio De Janeiro in June 1992 (Brysk, 2000: 130, 135, 231). They have also filed legal cases against governments, private companies, and other institutions. In addition, they have pushed for ethics statements to be crafted and have stipulated that individual researchers and development workers follow them.

We then go on to discuss the ways in which countries in Asia have handled protected areas and indigenous peoples. Asia is the part of the world with the largest number of indigenous people (Erni, 2008; Asia Indigenous Peoples Pact, 2014; Rights and Resources Initiative, 2022; Mamo, 2023).

We begin with an assessment of cases that brought up serious ethical issues about rights of indigenous peoples, including the Huaorani of Ecuador, who faced intense exploitation of their lands by multinational and national petroleum companies (see Rival, 2016). The Yanomami of Brazil and Venezuela have faced incursions of thousands of gold miners (*garimperos*) and groups seeking to exploit their timber and other resources (Watts, 2023). Indigenous South Americans have long been exploited in a variety of ways—from serving as laborers in mines, field hands on farms, loggers and rubber tappers in the Amazon Basin, and herders on settler ranches (see, for example, Davis, 1977; Brysk, 2000; Postero & Zamosc, 2006).

© The Author(s) 2023
M. Sapignoli, R. K. Hitchcock, *Anthropology and Ethics*,
https://doi.org/10.1007/978-3-031-39268-9_5

Table 5.1 Latin American countries and their indigenous peoples

Name of Country	Size (km²)	Population size (2022)	Indigenous peoples and percentage		No. of National Parks
Argentina	2,780,400	46,245,668	955,032	2%	35
Bolivia	1,098,581	12,054,379	6,016,026	51%	22
Brazil	8,515,770	217.240,060	896,917	4%	67
Chile	756,102	18,430,408	1,805,243	9.9%	41
Colombia	1,138,910	49,059,221	1,559,852	3%	51
Ecuador	283,561	17,289,534	1,018,076	6%	11
French Guiana	32,253	318,400	21,279	8.5%	1
Guyana	214,969	789,683	67,525	9%	2
Paraguay	406,752	7,356,409	112,848	1.5%	15
Peru	1,285,216	32,275,736	7,021,271	22%	14
Surinam	163,820	632,638	47,892	7.8%	13
Uruguay	176,215	3,497,213	76,452	2.2%	22
Venezuela	912,050	29,789,730	724,592	2.5%	43
Central America					
Belize	22,966	412,387	52,618	13%	17
Costa Rica	51,100	5,204,411	104,143	2%	28
El Salvador	21,041	6,568,745	14,408	2.2%	5
Guatemala	108,889	17,703,190	5,881,009	34%	31
Honduras	112,090	9,459,440	536,541	5.8%	19
Mexico	1,964,375	129,150,971	16,933,283	3%	67
Nicaragua	130,370	6,301,580	518,104	8.3%	2
Panama	75,420	4,337,668	417,559	10.7%	10
Total 21 countries	(km²)				Ca.

Note: Data obtained from the *World Factbook* 2023, and from CEPAL (2014). *Los Pueblos Indigenas en América Latina: Avances en el Ultimo Decenio y Retos Pendientes para la Garantia de sus Derechos*. Santiago: Chile: Comisión Económica para América Latina y el Caribe (CEPAL)

In the past several decades indigenous activism in South America has picked up substantially. The activism takes a number of forms: demonstrations against government policies, participating in strikes, seeking negotiations with governments, taking part in lawsuits, attending international meetings such as the United Nations Permanent Forum on Indigenous Issues (UNPFII), and seeking political office in the various countries of Central and South America (see Table 5.1) for a list of these countries and their indigenous populations. Some countries, including Bolivia, Ecuador, and Peru have experienced serious instability in recent years, in part because of indigenous dissatisfaction with their governments' policies (Mamo, 2023).

In central and South America, indigenous people have been affected substantially by the establishment of protected areas, where 7.1% of the sub-continent is under strict protection (Baldi et al., 2019). The degree to which indigenous peoples are affected by protected areas varies by country. There are still some areas of Amazonia

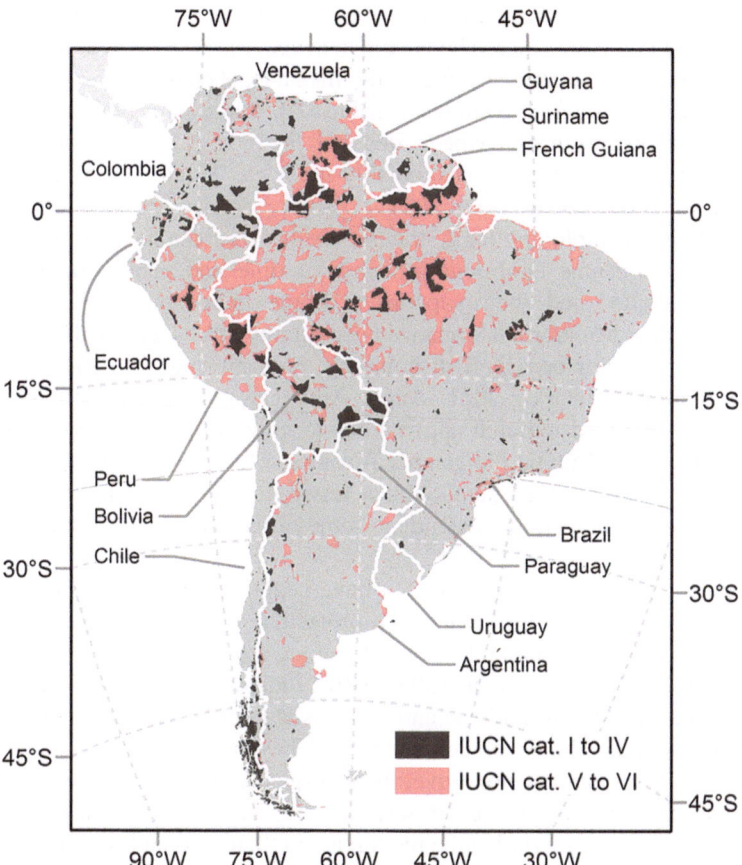

Fig. 5.1 Protected areas in South America

which lack protection in spite of the fact that there are biodiversity hotspots and voluntary isolated indigenous peoples (VIIPs) residing in them. Many of these areas have faced an expansion in agricultural development, settlement, and extractive industries. Deforestation rates have expanded in many areas, particularly under the government of Jair Bolsonaro, though the rates have slowed more recently, especially in those areas under indigenous peoples' protection. Indigenous lands in protected areas in the tropical zones of south America exhibit a high degree of conservation (Sze et al., 2022). High rates of conservation are also found in national parks such as that of Yasuni in Ecuador where the Huaorani and other indigenous groups reside (Rival, 2016). Figure 5.1 shows protected areas in South America.

An important strategy of South American indigenous peoples and their supporters has been to take governments and transnational companies to court. Sometimes this was done in order to seek compensation or to get relief from massive human rights violations, as seen, for example, with the Huaorani of the Oriente Region of

Ecuador, whose lands and resources have been destroyed by the actions of American and Ecuadorian oil companies. The Huaorani sought the assistance of the Rainforest Foundation in San Francisco and eventually filed a legal claim in Los Angeles under the Alien Claims Torts Act (as shown in Table 5.4). While some of these claims have been dismissed, others that have been filed with the Inter-American Court of Human Rights and have had some success, notably the Awas Tingni case in Nicaragua in 2001 (Anaya, 2009) and Belize in 2021.

A drawback to employing legal strategies is that they can be enormously expensive, and the consume enormous amounts of time and energy on the part of indigenous peoples and their supporters. Some indigenous activists, such as Alessandra Korap Munduruku, organized community efforts among the Munduruku to stop mining development by British mining company Anglo American in Brazil's Amazon rainforest. In May 2021, Anglo American committed itself formally to withdraw 27 approved research applications to mine inside Indigenous territories, including the Sawré Muybu Indigenous Territory, which is 438,000 acres in extent. For her efforts, she received the Goldman Environmental Prize in 2023. There are also risks to indigenous rights defenders, some of whom have been killed for their efforts.

Indigenous people in South America have negotiated with the nation-state and with oil companies in an effort to increase protection of forest resources, as seen, for example, with the Huaorani and other indigenous groups in the case of the Yasuní National Park in Ecuador (Rival, 1993, 2010). Indigenous South American groups have held meetings, consultations, and field visits in order to assess the impacts of petroleum and other kinds of development. They have collaborated with local, national, and international non-government organizations. They have also taken part in participatory mapping exercises, and they have met with legal experts to discuss how best they might handle the threats to their land and resources.

In many ways, indigenous peoples in South America have had a greater degree of success in establishing some control over protected areas, some of which have been declared Indigenous Protected and Conservation Areas (IPCAs). One of the reasons for this situation is that indigenous people have had the support of international organizations such as Survival International, Cultural Survival, and the International work Group for Indigenous Affairs. Some South American countries such as Brazil have sought to set aside protected areas in the Amazon Basin aimed at conserving both tropical forests and the people who reside in them. Asmittedly, the degree to which indigenous peoples' rights have been protected in Brazil have varied over time. During the presidency of Jair Bolsonaro (2019–2022), for example, areas set aside for indigenous people were reduced in number, and invasions by farmers, miners, and loggers occurred. International organizations such as the United Nations and the World Bank have sought to reinforce conservation protections and to enhance protections of indigenous lands.

Asia Indigenous Peoples and Protected Areas

Asia, the world's largest continent, stretches from India to north China and south to Indonesia, Malaysia and west to the Philippines and Japan, has the highest number of indigenous people in the world (see Table 5.2). There are some 1300 national parks in Asia according to the World Protected Areas Data Base. While the majority of indigenous peoples live in diverse settings in forests, mountains, savannas, and on islands, they generally do not have rights in Asia's protected areas (Colchester & Erni, 1999; Rights to Resources Initiative, 2022).

In Asia, examples of indigenous peoples living within traditional lands that have been set aside as national parks include the Kayan Mentarang National Park in East Kalimantan, Indonesia, Crocker Range Park in Sabah, Malaysia; Northern Sierra

Table 5.2 Population sizes and numbers of national parks in Asian countries

Country	Area in sq km	Population size (2023)	No. of National Parks
Afghanistan	652,230	39,232,002	8
Bangladesh	148,460	167,184,465	8
Bhutan	38,394	876,181	5
Burma	676,578	57,970,293	9
Cambodia	181,035	16,891,245	7
China	9,596,960	1,413,142,846	225
India	3,287,263	1,399,179,585	106
Indonesia	1,904,599	279,476,346	54
Japan	377,915	123,719,238	34
Korea North	120,538	26,072,217	9
Korea South	99,720	51,966,948	22
Laos	236,800	7,852,377	20
Malaysia	329,847	34,219,975	25
Mongolia	1,564,116	3,255,468	24
Nepal	147,181	30,899,443	12
Pakistan	796,095	247,653,551	34
Philippines	300,000	116,434,200	35
Russia	17,098,248	141,698,922	64
Singapore	719.2	5,975,383	1
Sri Lanka	65,610	23,326,272	26
Taiwan	35,980	23,588,613	9
Thailand	513,120	69,794,997	47
Vietnam	331,210	104,799,174	30
25 countries			

Note: Data obtained from *The World Factbook* (www.cia.gov, accessed 10 May 2023, the IUCN, the world protected area database, and from government reports and censuses, work of researchers, development agencies, national archives, government and international agency reports, and documents of non-government organizations including Minority Rights Group International, the International Work Group for Indigenous Affairs, Survival International, the Forest Peoples Program, the Asia Indigenous Peoples Pact, and fieldwork

Source: Wildlife Protection Society of India

Fig. 5.2   Tiger Reserves in India

Madre National Park in Mindanao, Philippines where several hundred Agta reside; Virachey National Park in Ratanakiri, Cambodia; Royal Chitwan National Park in Nepal; Ayubia National Park in Pakistan; and the Rajaji National Park in northwest India. In the case of India, there are over 53 tiger reserves in the country, most of which do not allow indigenous and tribal people to reside inside the boundaries (Rushkin, 2023). Figure 5.2 shows the distribution of tiger reserves in India.

Only three of Asia's countries meet the criteria of 17% of their land devoted to terrestrial protected areas: Bhutan, Nepal, and Sri Lanka (Chowdhury et al., 2021). Most of Asia's protected areas are relatively small in size (Juffe-Bignoli et al., 2014). Many of these protected areas face a complex array of threats, both natural and anthropogenic. Population growth on the peripheries of protected areas is a major problem, as is the pressure to allow extractive industries to take place within the boundaries of protected areas.

There are relatively few national parks in Asia where indigenous people have been allowed to continue to reside. One of these is Bukit Duabeles National Park in Indonesia which is 605 km^2 in size. There are some 500 Orang Rimba residing inside of this national park (Elkholy, 2016; Ramsey Elkholy personal communication, January 2023). There are at least a few hundred Karen residing in one of Thailand's largest national parks, Kaeng Krachan National Park, which is 2914 km^2 in size. In Japan, on the other hand, the Ainu have been excluded from the Akan-Mahu National Park, established in 1934, which is 914.13 km^2 in size. Instead, the Ainu live in a township on the peripheries of the national park and operate a cultural museum there. The Ainu see the park area as being an important part of their ancestral territory which contains sacred ritual and cultural sites. In the case of Taiwan, few of the 13 aboriginal groups have been allowed to remain in national parks, but they do have rights to collect resources in some of them (Pei-Lin Yu, personal communication, December 18, 2022). In 2021, the High Court of Taiwan upheld restrictions on hunting which upset indigenous groups (Chien & Qin, 2021).

Asian countries that have been involved in the construction of dams have sought in some cases to establish protected areas in the vicinities of dams (see Table 5.3). An example of such protected areas related to large dams is that of Nam Theun 2 Dam in Laos where indigenous groups have been given rights to continue to fish and exploit wild plant resources (Scudder, 2019a, b). The Mahaweli Water Project in Sri Lanka also saw the establishment of protected areas associated with the project, but the Wannialetto have been excluded from these areas (Ted Scudder, personal communication, 2022). The long-term impacts of resettlement of people from dams and protected area establishment have been significant (Fujikura & Nakayama, 2013; Scudder, 2019a). In some cases, the construction of dams and setting aside of land for the reservoirs have resulted in a rise in indigenous activism. Land dispossession has resulted in a loss of indigenous culture, livelihoods, and indigenous knowledge.

Some of the strategies employed by indigenous peoples have been legal in nature. Indigenous peoples have sought the services of lawyers to work with them in taking governments to court in order to establish their land and resource rights, as has occurred, for example, in the cases of Belize and Nicaragua (see Table 5.4). A number of these cases saw collaborations among indigenous peoples, lawyers, and anthropologists. These cases often involve extensive financial investment which is often beyond the means of local indigenous communities. Fortunately for indigenous peoples, there are a number of support organizations which they have been able to call upon to assist them, including Survival International, the International Work Group for Indigenous Affairs, and the Forest Peoples Program. They have also been able to access funds for their activities from the European Union and various European governments in addition to the World Ban and the United Nations.

Awareness-raising by indigenous peoples and their supporters has occurred at international meetings, including at the United Nations Permanent Forum on

Table 5.3 Major dams in Asia that have affected local communities and that are associated with protected areas

Name of Dam and Year(s) of Completion	River and country	Numbers of households of individuals displaced or resettled
Arun III	Arun River, Nepal	775 people displaced, reservoir 43 hectares in size
Bakun Dam, 2000	Balul River, Borneo, Malaysia, mostly state funded	Toxic impacts, removals of Penan hunter-gatherers
Dahla (Arghandab) Dam, 1954, additions, 2012–2018	Arghandab River, Kandahar, Afghanistan	Estimated 25 households were resettled as a result of the heightening of the dam wall
Nam Ngum Dam, 1972	Lao People's Democratic Republic (Laos), state funded	3500 people displaced by the project, 37,000 ha reservoir; fishing concession to a private entrepreneur
Nam Theun 2 (NT2) dam, commissioned 2010, water from Nam Theun River released into Xe Bang Fai River	Lao People's Democratic Republic (Laos); World Bank and Asian Development Bank and private funding, one of the largest internationally financed project in Asia	10,000 people resettled, some of them hunter-gatherers; 40,000–150,000 total project-affected people; conservation areas were developed as part of the project, 450 km^2 (170 m^2) reservoir
Nam Theun-Hinboun hydro-power project	Nan Theun Hinboun River, Lao People's Democratic Republic (Laos)	630 ha reservoir, 6000 people from 25 villages resettled, some of them indigenous
Sardar Sarovar Dam, 1987–2017, second largest concrete megadam in the world	Narmada River, India	Estimated 100,000 people resettled, 140,000 people affected by infrastructure, canals, irrigation systems, 90,820 ha reservoir
Three Gorges Dam, China, 2006, powerplant 2012	Yangtze River	1.3 million people impacted, 600 towns, 1084 km^2 (419 sq. mi) reservoir

Note: Data obtained from fieldwork and from the International Commission on Large Dams; Scudder (2005:59–60, 2019:1–9); http://www.dams.org and International Rivers http://www.internationalrivers.org, accessed 3 April 2023

Indigenous Issues annual meetings (see Table 5.5). While the rules of the Permanent Forum are such that indigenous representatives cannot take formal issue with the governments of the states where they reside, the UNPFII does allow for meetings between indigenous peoples and non-government organizations, and they can also learn about lawyers and barrister who may be willing to help them in their legal work. Fieldwork at the United Nations has been useful in showing the ways in which indigenous peoples can learn from each other about strategies to be employed in handling legal matters.

Table 5.4 Legal cases involving land and resource rights of indigenous peoples

Group	Locality	Country	Issue	Workers
Aboriginals and Torres Strait Islanders (the Mabo Case, 3 June, 1992)	Mer Island, Murray Islands, Torres Straits	Australia	Land rights	Mabo family, Barbara Hocking (lawyer), Nonie Sharp (anthropologist)
Ke'kchi and Mopan Maya	Southern Belize	Belize	Land and resource rights	S. James Anaya (lawyer), Laurie Medina (anthropologist)
G/ui, G//ana, Bakgalagadi (High Court Case, Lobatse, (13 December, 2006)	Central Kalahari Game Reserve and Ranyane	Botswana	Land rights, subsistence hunting rights	John Whitehead Gordon Bennett (lawyers), George Silberbauer, Robert Hitchcock, (anthropologists)
Delgamuukw Gitxsan and Wet'suwet'en tribes of British Columbia, Canada Supreme Court 1997	British Colombia	Canada	Land rights	Canada supreme court lawyer, Peter Bennett, UBC historian and several anthropologist including Richard Daly
Baka Batwa vs Worldwide Fund for Nature (WWF) and Ministry of Forests and Fauna for human rights violations, dispossession, 2014–2016	Southeastern Cameroon	Cameroon	Formal complaint to OECD and Ministry of Forests and Fauna by Survival International, 10 February, 2016, dropped 2018	OECD (Organization for Economic Development and Cooperation in Europe) and government of Cameroon, Gordon Bennett (lawyer)
Huaorani Alien Claims Tort Act case	Oriente Region	Ecuador	Free, prior, and informed consent Land and Resources Rights and Inter-American Court on Human Rights	Ecuador lawyers, Laura Rival (anthropologist)
Sarayaku People 2012	Sarayaku People	Ecuador and Compania General de Combustibles oil company	Land and Resources and environmental rights Inter-American Court of Human Rights	Inter-American Court lawyer, Karla Quinana (anthropologist)
Jawara, Onge 2002	Andaman Islands	India	Land rights, removal of settlers from tribal reserves by High Court	Gordon Bennett (lawyer), V. Venkatswar (anthropologist)

(continued)

Table 5.4   (continued)

Group	Locality	Country	Issue	Workers
Ainu people Nibutani Dam case *Kayano v. Hokkaido Expropriation Committee* 1997	Sapporo District	Japan	Cultural and Land Rights Sapporo Court	Sapporo Court lawyer, T. Irimoto (anthropologist), K. Tekuza (anthropologist)
Endorois, 2009	Rift Valley	Kenya	Land and resource rights African Commission for Human and Peoples Rights (2010)	Minority Rights Group lawyer, K. Sing'Oei anthropologist Justin Kenrick (anthropologist)
Temuan people, 2005	Bukit Tampoi	Malaysia	Land and Resources Rights, compensation Malaysia's Court of Appeal	Malaysia Court of Appeal Lawyer, George Appell (anthropologist)
Khwe San 1998	Popa Falls Prison Farm, Okavango River	Namibia	Land rights (settled out of court)	Namibia Legal Assistance Center lawyer (Peter Watson) Gertrud Boden (anthropologist) IRDNC staff, WIMSA staff
Hai//om San 2015	Etosha National Park resettlement case	Namibia	Land rights dismissed by High Court, appeal filed	Namibia Legal Assistance Center lawyers (Peter Watson, Willem Odendaal) Ute Dieckmann, Thomas Widlok (anthropologists)
Mayagna (Sumo), 2001	Awas Tingni Community, Atlantic Coast	Nicaragua	Inter-American Court of Human Rights Land and resource (forestry) rights	S. James Anaya lawyer, P. Grossman (anthropologist)
Xákmok Kásek Indigenous Community 2010	Xákmok Kásek	Paraguay	Inter-American Court of Human Rights	IACHR lawyer, Alejandro Peralada (anthropologist)
Northern Aché people, Paraguay	Aché people	Paraguay	Case 7300/2013 was opened by Argentine Federal Court No. 5	Argentine lawyer, Alejandro Peralada (anthropologist)

(continued)

Table 5.4 (continued)

Group	Locality	Country	Issue	Workers
‡Khomani San 1998	Kgalagadi Transfrontier Park	South Africa	Land rights, co-management rights	Roger Chennels (lawyer), Hugh Brody, Nigel Crawhall (anthropologists)
Nama *Richtersveld Community vs Alexkor Ltd* (Constitutional Court, October, 2003)	Nama people, Richtersveld National Park	South Africa	Mineral rights, grazing rights	Legal Resources Foundation (Henk Smith) lawyer; Emile Boonzaier (anthropologist)
Saramaka people 2007 Inter-American Court of Human Rights	Saramaka People	Suriname	Land and Resources Rights, FPIC	Inter-American Court of Human Rights lawyer, anthropologist unknown
Mashpee Tribe versus Mashpee Town (1976–1978)	Mashpee Wampanoag of Cape Cod	Massachusetts, United States	Right of recognition, land and resource rights	US. Federal court, Mashpee Lawyer, and anthropologists

Conclusions

In conclusion, strategies for addressing the challenges of protected area declaration employed by indigenous people include negotiation, demonstrations, collaboration with park managers and state institutions, establishment of indigenous protected areas, working out co-management systems, and arranging power-sharing agreements. Going to court, while in many ways an effective strategy, is an expensive and time-consuming one. Grassroot resistance has proven to be effective in many parts of the world (Oliver-Smith, 2010). The majority of indigenous people who have been involuntarily resettled as a result of the establishment of protected areas would prefer to work cooperatively with park administrators, scientists, and nation-states in order to ensure that the benefits of protected areas are available to all. The key term for many of the world's indigenous people is 'conservation with justice.'

Table 5.5 Meetings of the United Nations Permanent Forum on Indigenous Issues (UNPFII)

Year	Date	Session	Special theme	Region of half day discussion
2002	12–24 May	First	–	–
2003	11–23 May	Second	Indigenous children and youth	–
2004	10–21 May	Third	Indigenous women	–
2005	16–27 May	Fourth	MDGs and indigenous peoples with focus on Goal 1 to Eradicate Poverty and Extreme Hunger, and Goal 2 to achieve universal primary education	–
2006	15–26 May	Fifth	The MDGs and indigenous peoples: re-defining the MDGs	Africa
2007	14–25 May	Sixth	Territories, Land and Natural Resources	Asia
2008	21 April–2 May	Seventh	Climate Change, bio-cultural diversity and livelihoods: the stewardship role of indigenous peoples and new challenges	Pacific
2009	18–29 May	Eighth	Review Year	Artic
2010	19–30 April	Ninth	Indigenous peoples: development with culture and identity, art. 3 and art. 32 of the UNDRIP	North America
2011	16–27 May	Tenth	Review Year	Central and South America and the Caribbean
2012	7–18 May	Eleventh	The Doctrine of Discovery and its enduring impact on indigenous peoples and the right to redress for past issues, Article 28 and Article 37 of the UNDRIP	Central and Eastern Europe, the Russian Federation, Central Asia and Transcaucasia
2013	20–31 May	Twelfth	Review Year	Africa
2014	12–23 May	Thirteenth	Principles of good governance consistent with the United Nations Declaration on the Rights of Indigenous Peoples: Articles 3 to 6 and 46	World Conference on Indigenous Peoples
2015	20 April–1 May	Fourteenth	Indigenous Peoples and Post-2015 Development Agenda, with a focus on Hunger and Disease	The Pacific Region
2016	9–20 May 2016	Fifteenth	Indigenous peoples: conflict, peace and resolution	Worldwide

(continued)

Table 5.5 (continued)

Year	Date	Session	Special theme	Region of half day discussion
2017	24 April–5 May 2017	Sixteenth	Tenth Anniversary of the United Nations Declaration on the Rights of Indigenous Peoples: Measures taken to implement the Declaration	Worldwide
2018	16–27 April 2018	Seventeenth	Discusions of Conservation and Indigenous Peoples, Land Rights, and the Sustainable Development Goals	Land and resources
2019	22 April–3 May 2019	Eighteenth	Traditional knowledge: Generation, transmission and protection	Worldwide
2020	12–24 April 2020	Nineteenth	postponed	
2021	19–30 April 2021	Twentieth	Peace, justice and strong institutions: the role of indigenous peoples in implementing Sustainable Development Goal 16, Conducted by Zoom	Worldwide
2022	25 April–6 May 2022	Twenty First	Indigenous peoples, business, autonomy and the human rights principles of due diligence including free, prior and informed consent"	Worldwide
2023	17–28 April 2023	Twenty Second	Indigenous Peoples, human health, planetary and territorial health and climate change: a rights-based approach	Worldwide

References

Anaya, S. J. (2009). *Indigenous peoples in international law*. Wolters Kluwer.

Asia Indigenous Peoples Pact. (2014). *Overview of the state of indigenous peoples in Asia*. Asia Indigenous Peoples Pact.

Baldi, G., Schauman, S., Texeira, M., Marinaro, S., Martin, O. A., Gandini, P., & Jobbágy, E. G. (2019). Nature representation in South American protected areas: country contrasts and conservation priorities. *PeerJ, 7*, 1–23. https://doi.org/10.7717/peerj.7155

Brysk, A. (2000). *From tribal village to global village: Indian Rights and International Relations in Latin America*. Stanford University Press.

Chapin, M., Lamb, Z., & Threlkeld, B. (2005). Mapping indigenous lands. *Annual Review of Anthropology, 34*, 619–638.

Chapin, M., & Threlkeld, B. (2001). *Indigenous landscapes: A study in ethnocartography*. Center for the Support of Native Lands.

Chien, A. C., & Qin, A. (2021). Top Court in Taiwan upholdes restrictions on Hungin, Angering indigenous Groups. *New York Times*, 8 May 2021.

Chowdhury, S., Alam, S., Labi, M. M., Nahla Khan, M., Rokonuzzaman, D. B., Tahea, T., Mukul, S. A., & Fuller, R. A. (2021). Protected areas in South Asia: Status and prospects. *Science of the Total Environment, 811*. https://doi.org/10.1016/j.scitotenv.2021.152316

Colchester, M., & Erni, C. (Eds.). (1999). *Indigenous peoples and protected areas in South and Southeast Asia*. International Work Group for Indigens Affairs.

Davis, S. H. (1977). *Victims of the miracle: Development and the Indians of Brazil*. Cambridge University Press.

Elkholy, R. (2016). *Being and becoming: Embodiment and experience among the Orang Rimba of Sumatra*. Berghahn Books.

Erni, C. (2008). *The concept of indigenous peoples in Asia: A resource book*. International Work Group for Indigenous Affairs and Asia Indigenous Peoples Pact Foundation.

Fujikura, R., & Nakayama, M. (2013). The long-term impacts of resettlement programmes resulting from dam construction projects in Indonesia, Japan, Laos, Sri Lanka and Turkey: a comparison of land-for-land and cash compensation. *International Journal of Water Resources Development, 29*(1), 4–13.

Herlihy, P. H., & Knapp, G. (2003). Maps of, by, and for the peoples of Latin America. *Human Organization, 62*(4), 303–314.

Juffe-Bignoli, D., Bhatt, S., Park, S., Eassom, A., Belle, E. M. S., Murti, R., Buyck, C., Raza Rizvi, A., Rao, M., Lewis, E., MacSharry, B., & Kingston, N. (2014). *Asia protected planet 2014: Tracking progress towards targets for protected areas in Asia*. UNEP-WCMC.

Mamo, D. (Ed.). (2023). *The indigenous world 2023*. International Work Group for Indigenous Affairs.

Oliver-Smith, A. (2010). *Defying displacement: Grassroots resistance and the critique of development*. University of Texas Press.

Poole, P. (1995). *Indigenous peoples, mapping, and biodiversity conservation: An analysis of current activities and opportunities for applying geomatics technologies*. BSP Peoples and Forest Program Discussion Paper. Washington, D.C.: Biodiversity Support Program.

Postero, N. G., & Zamosc, L. (Eds.). (2006). *The struggle for indigenous rights in Latin America*. Sussex Academic Press.

Rights and Resources Initiative. (2022). *Reconciling conservation and global biodiversity goals with community land rights in Asia*. Rights and Resources Initiative (RRI).

Rival, L. (1993). Confronting petroleum development in the Ecuadorian Amazon: The Huaorani, human rights and environmental protection. *Anthropology in Action (BASAPP), 16*, 14–15.

Rival, L. (2010). Ecuador's Yasuní-ITT initiative: The old and the new values of petroleum. *Ecological Economics, 70*, 358–365.

Rival, L. (2016). *Huaorani transformations in 21st century Ecuador. Treks to the future of time*. University of Arizona Press.

Rushkin, E. (2023). *The Nature of Endangerment in India: Tigers. 'Tribes,' Extermination, and Conservation 1818–2020*. Oxford University Press.

Scudder, T. (2019a). *Large dams, long-term impacts on riverine communities and free flowing rivers*. Springer.

Scudder, T. (2019b). A retrospective analysis of Laos's Nam Theun 2 Dam. *International Journal of Water Resources Development*. https://doi.org/10.1080/07900627.2019.1677456

Sze, J. S., Childs, D. Z., Roman Carrasco, L., & Edwards, D. P. (2022). Indigenous lands in protected areas have high forest integrity across the tropics. *Current Biology, 2022*, 4949–4956. e3. https://doi.org/10.1016/j.cub.2022.09.040

Watts, J. (2023). Health emergency over Brazil's Yanomami. *Lancet, 401*, 631.

Chapter 6
Conservation, Ethics, and Indigenous Peoples

Introduction

In this chapter, we assess conservation, ethics, and indigenous peoples. Ethics have become increasingly a focus of discussion and debate in conservation science, anthropology, and indigenous peoples' rights in the past several decades. Part of the reason for this is that there have been complaints from indigenous peoples about how they have been treated by conservationists and conservation organizations, as seen, for example in central and southern Africa. Indigenous peoples broadly want to see a rights-based approach (RBA) to conservation. They want to ensure that their human rights are respected when it comes to be establishment and running of protected areas. Global environmental ethics are seen as being crucial by both scholars and practitioners (Pojman, 2000).

The disciplines of conservation science and anthropology have had a complicated history when it comes to human rights, and especially indigenous peoples' rights. While anthropologists were largely supportive of indigenous peoples and their rights in the in the nineteenth and twentieth centuries, they were also sometimes blamed for the exploitation of indigenous peoples' traditional and cultural knowledge and information. Anthropologists like Lewis Henry Morgan, who was also a lawyer, stood up for American Indians on the Great Plains and elsewhere. James Mooney of the Smithsonian Institution sought to get the War Department of the United States not to consider the Ghost Dance as a 'war dance,' in the 1890s, but the U.S. 7th Cavalry took revenge on Lakota engaged in the what the government suggested was the Ghost Dance at Wounded Knee, South Dakota on 27 December 1890. This event, which brought to an end the so-called 'Indian Wars' of the nineteenth century, led to the encapsulation of Indian communities on reservations and the efforts to acculturate indigenous people in the United States.

Numerous anthropologists worked on behalf of American Indians during the Indian Claims Commission (ICC) (1946–1971). Anthropologists came into some disrepute because of the refusal of the Executive Board of the American

© The Author(s) 2023   97
M. Sapignoli, R. K. Hitchcock, *Anthropology and Ethics*,
https://doi.org/10.1007/978-3-031-39268-9_6

Anthropological Association (AAA) to support the draft of the Universal Declaration on Human Rights in 1947 (Engle, 2001). This led to a split in the AAA and the establishment of the Society for Applied Anthropology (SfAA). More recently, anthropologists helped found a Committee for Human Rights (CfHR) in the AAA, which unfortunately was done away with in the past several years.

Anthropologists have been supportive of indigenous social movements such as those opposing the Keystone XL Pipeline, the Dakota Access Pipeline DAPL in North Dakota, and the Line-3 Pipeline in Minnesota. And during the Trump Administration (2017–2021) anthropologists worked with various indigenous groups to oppose policies that were harmful to native peoples. During the COVID-19 pandemic starting in February 2021, anthropologists have been working with indigenous peoples throughout the country to help provide information, personal protective equipment (PPE), face masks, hand sanitizer, soap and food.

The interactions between anthropologists and the civil and criminal justice systems in nation-states have a long history, much of which is marked by strident debate and criticism as well as considerable accomplishment (Engle, 2001, 2010; Fluehr-Lobban, 2013). Anthropological expertise was called into question by courts in the United States, Canada, Australia, and South Africa. Ethics questions were raised about the testimony of anthropologists, who were said to be overly partisan with respect to the people with whom they worked. Statements by anthropologists in cases ranging from the Central Kalahari in Botswana to indigenous land rights cases in western Canada were challenged as being overly emotional and even 'historically and culturally inaccurate.'

The Code of Ethics of the Anthropological Association, like nearly all ethics codes, calls for 'Do no harm' (American Anthropological Association, 2012). It also calls for the protection of the identity of informants, something that has been watered down over the years, in part because of what occurred on 11 September 2001. Anthropologists, like other social scientists, believe that they have a social responsibility, including to those people with whom they work.

As John Hardwig (1994: 78) says:

> The relationship between expert and layperson is grounded on an epistemic inequality. The expert knows more than the layperson about matters within the scope of her expertise. And if the layperson appeals to the judgment of the expert, he usually does so because he acknowledges the superiority of the expert's judgment to his own. Thus, the epistemology of the expert-layperson relationship can be focused on the concept of rational deference to epistemic authority. This rational deference lies at the heart of the particular form of power that an expert has and is also the center of the particular form of vulnerability that each of us, as a layperson, is in.

In most cases, anthropologists lack power, especially in the countries where they work. All too often, anthropologists who challenge governmental authorities may find themselves declared as prohibited immigrants, as occurred in Botswana, Ethiopia, and Guatemala in the past several decades. While their expertise may lie in extensive knowledge of a society, that knowledge may appear to be a risk to a government that is devoting its energy to large-scale development, some of which leads to local-level displacement, something that anthropologists often disagree with

(Engle, 2010). Sometimes what anthropologists argue is direct, first-person evidence is seen as 'hearsay evidence' in a court of law. One way to get around this problem is for the lawyer to collaborate closely with the anthropologist in preparing her or his testimony ahead of time. Anthropologists should not argue that they are 'speaking for the community' since they are usually not part of that community; they are outsiders, albeit ones who have spent extensive time with specific communities and may know the language, culture, and history of that community well. They should admit when they do not know something when they are in court. They should also tell the truth at all times—'the truth, the whole truth, and nothing but the truth.' As an expert, one needs to consider the impacts of any statements made not just on the people who they are testifying for in court, but all people affected by those statements.

Some of the strategies employed by indigenous people are risky, taking governments and transnational corporations to court, for example given the costs involved and there are substantial chances for the case being lost or dismissed. Engaging in uprisings, such as occurred among the Lacandon Maya in Chiapas, Mexico in 1994 or the Shining Path efforts in Peru in the 1970s and 1980s, can also be seen as risky due to state responses to the uprisings.

Anthropologists have generally been supportive of indigenous peoples' human rights efforts, particularly in North and South America and Africa. Anthropological ethics statements have been refined over time in order to take into consideration the perspectives of the people with whom they work. The ethics principles such as 'do no harm' and ensuring that research and field work is transparent and open to assessment by the communities and that benefits accrue to project stakeholders are important, Indigenous peoples increasingly are seeking to come up with ethics statements and are holding anthropologists and other social scientists to account. International organizations such as the United Nations and the World Bank have social safeguards policies that seek to protect local people involved in projects. Indigenous peoples and their support organizations along with nation-states and international organizations participate regularly in the annual meetings of the United Nations Permanent Forum on Indigenous Issues in New York.

The Universal Declaration on the Rights of Indigenous Peoples (United Nations, 2007), as noted by Professor James Anaya (2023), contains five interrelated ideals: equality, cultural survival, communal as well as individual property, self-determination, and economic and social justice. The declaration is a key example of the power of ideas that indigenous people have brought to the international community. Anthropologists and indigenous people are committed more and more to principles of social responsibility and justice (Fluehr-Lobban, 2013; Anaya, 2009), and these principles are crucial to the future survival both of indigenous people and the discipline of anthropology.

The Society for Conservation Biology (SCB), the Wilderness Society, and other conservation societies and organizations have developed ethics states in the past two decades that seek to ensure responsible care for the earth and its ecosystems. They seek to ensure that human subjects (i.e., those affected by conservation programs) are treated with dignity and respect. (See, for example, the code of ethics of the

Society for Conservation Biology, 2004: 2). All of the major conservation non-government organizations, including the World Wildlife Fund, the Worldwide Fund for Nature, the Nature Conservancy, and Conservation International, have developed ethics statements that seek to protect the rights of indigenous peoples.

Some of the conclusions of conservation organizations include the following:

- Large conservation projects should promote integration of biodiversity conservation and sustainable use to deliver a number of long-term socio-economic benefits to communities involved. Examples include the creation of locally managed institutional mechanisms including community development funds and conservation stewardship programs, community forestry projects that benefit people living around them, subsidies to local communities to develop alternative fuel sources, capacity building of local organizations, educational courses and the development of awareness raising programmes (see Juffe-Bignoli et al., 2014: 48).
- Evaluation of the effectiveness and appropriateness of protected area networks is a vital component of responsive, pro-active protected area management for connectivity which is an objective of conservation efforts (see Juffe-Bignoli et al., 2014: 48).
- Projects should develop indicators to assess the ecological health of ecosystems and the sustainability of supporting institutions which provide an important measure of the success of conservation initiatives at large scales.

Drawing on the lessons from the experiences of indigenous peoples and from the organizations with whom they have interacted, it is possible to come up with a set of recommendations about best practices that could be employed in future work vis a vis protected areas. These are listed below.

- There should be no removals of indigenous peoples from protected areas without Free, Prior, and Informed Consent (FPIC) having been obtained.
- Arrangements should be made for alternative land along with fair and just cash compensation for people removed from protected areas.
- Co-management systems must be set up for protected areas that involve all stakeholders, giving priority to indigenous peoples and others who are resettled.
- Mapping should be done of indigenous peoples' territories inside of protected areas and indigenous peoples should be able to participate in this.
- Indigenous peoples should have rights of access to protected areas for purposes of visiting sacred sites.
- Benefit-sharing arrangements should be established to ensure that indigenous peoples get a portion of the gate receipts of protected areas.
- Extractive industries should be kept to a minimum in protected areas and only done after Free, Prior, and Informed Consent is obtained.
- Indigenous peoples have a say in the ways in which extractive industries are pursued, and the companies engaged in these activities should employ systems of Corporate Social Responsibility (CSR).

- Indigenous peoples should also get direct benefits from the extractive activities of companies operating in protected areas.
- Land titling should be done for indigenous peoples' areas on the peripheries of protected areas.
- There should be no impunity for park authorities, militaries, or conservations who torture, murder, or otherwise mistreat indigenous and other peoples in the process of law enforcement and relocation activities.
- There should be a balance of power among protected area authorities, scientists, and indigenous peoples.

In conclusion, strategies for addressing the challenges of protected area declaration and management should ensure full-scale participation of project-affected people, including indigenous people. The strategies employed in protected areas should aim for sustainability, biodiversity conservation, and the promotion of wellbeing of all communities associated with protected areas. Equity and justice are key to the success of conservation and development efforts (Greiber et al., 2009; Martin, 2017: 161–164). Conservation should be both compassionate and attuned to the wellbeing of all species including humans, and ethics should be seen as a key aspect of conservation and anthropological science.

References

American Anthropological Association. (2012). *Statement on ethics: Principles of professional responsibilities*. American Anthropological Association.

Anaya, S. J. (2009). *Indigenous peoples in international law*. Wolters Kluwer.

Anaya, S. J. (2023). *Ideas about human rights and indigenous people*. Paper presented at the University of New Mexico, Albuquerque. New Mexico, 24 April 2023.

Engle, K. (2001). From skepticism to embrace: Human rights and the American Anthropological Association 1947-1999. *Human Rights Quarterly, 23*, 536–599.

Engle, K. (2010). *The elusive promise of indigenous development: Rights, culture, strategy*. Duke University Press.

Fluehr-Lobban, C. (2013). *Ethics and anthropology: Ideas and practice*. AltaMira Press.

Greiber, T., Janki, M., Orellana, M., Ssavaresi, A., & Shelton, D. (2009). *Conservation with justice*. International Union for the Conservation of Nature and Natural Resources (IUCN).

Hardwig, J. (1994). Towards an ethics of expertise. In D. Wueste (Ed.), *Professional ethics and social responsibility* (pp. 120–134). Rowman and Littlefield.

Juffe-Bignoli, D., Bhatt, S., Park, S., Eassom, A., Belle, E. M. S., Murti, R., Buyck, C., Raza Rizvi, A., Rao, M., Lewis, E., MacSharry, B., & Kingston, N. (2014). *Asia protected planet 2014: Tracking progress towards targets for protected areas in Asia*. UNEP-WCMC.

Martin, A. (2017). *Just conservation: Biodiversity, wellbeing and sustainability*. Earthscan, Routledge.

Pojman, L. P. (2000). *Global environmental ethics*. Mountain View, California and Mayfield Publishing Company.

Society for Conservation Biology. (2004). *Code of ethics*. Society for Conservation Biology.

United Nations. (2007). *Declaration on the rights of indigenous peoples*. United Nations.